Measuring the Natural Environment

Measurements of natural phenomena are vital for any type of
environmental monitoring, from the practical day-to-day
management of rivers and agriculture, and weather forecasting,
through to longer-term assessment of climate change and glacial
retreat. This book looks at past, present and future measurement
techniques, describing the operation of the instruments used and the
quality and accuracy of the data they produce.

The book describes the methods used to measure all the variables
of the natural world: solar and terrestrial radiation, air and ground
temperature, humidity, evaporation and transpiration, wind speed
and direction, rainfall, snowfall, snow depth, barometric pressue, soil
moisture and soil tension, groundwater, river level and flow, water
quality, sea level, sea-surface temperature, ocean currents and waves,
polar ice.

Measuring the Natural Environment is the first book to make a
thorough enquiry into the origins of environmental data, upon which
our scientific understanding and economic planning of the
environment directly hang. The book will be important for all those
who use or collect such data, whether for pure research or day-to-day
management. It will be useful for students and professionals working
in a wide range of environmental science: meteorology, climatology,
hydrology, water resources, oceanography, civil engineering,
agriculture, forestry, glaciology and ecology.

Ian Strangeways is Director of TerraData, a consultancy in
meteorological and hydrological instrumentation and data collection.
From 1964–89 he was Head of the Instrument and Applied Physics
sections at the Institute of Hydrology (Natural Environment
Research Council).

MEASURING THE NATURAL ENVIRONMENT

IAN STRANGEWAYS

CAMBRIDGE
UNIVERSITY PRESS

PUBLISHED BY THE PRESS SYNDICATE OF THE UNIVERSITY OF CAMBRIDGE
The Pitt Building, Trumpington Street, Cambridge, United Kingdom

CAMBRIDGE UNIVERSITY PRESS
The Edinburgh Building, Cambridge CB2 2RU, UK   http://www.cup.cam.ac.uk
40 West 20th Street, New York, NY 10011-4211, USA   http://www.cup.org
10 Stamford Road, Oakleigh, Melbourne 3166, Australia
Ruiz de Alarcón 13, 28014 Madrid, Spain

First published 2000

Printed in the United Kingdom at the University Press, Cambridge

Typeface 11/14 pt Times

A catalogue record for this book is available from the British Library

Library of Congress Cataloguing in Publication data

Strangeways, Ian, 1932–
    Measuring the natural environment / Ian Strangeways.
        p.   cm.
    Includes index.
    ISBN 0 521 57310 6
    1. Earth sciences – Measurement.   2. Environmental monitoring.
    I. Title.
    QE33.S79   2000
    363.7'063 – dc21   99-18381 CIP

ISBN 0 521 57310 6 hardback

QE33
.S79
2000

Contents

Acknowledgements

My experience of measuring the natural environment extends from 1964 to 1989 at the Institute of Hydrology (IH), first as head of the Instrument Section and later of Applied Physics, continuing after 1989 as consultant until the present. During this 35-year period, my work has been a mix of new instrument development and the application of existing equipment, new and old, to a variety of projects, many overseas and embracing all of the world's climates. Despite this, experience of every aspect of the subject has not been equal and I felt it advisable to check out some specialised areas and details that I was not entirely certain about. I would, therefore, like to acknowledge the advice of those listed below who helped in filling in the gaps and correcting my errors, producing a more reliable book. If there are remaining errors, it is not their fault but mine, for I am very aware of the ambitiousness of one person's attempt to write a book that covers so many fields. I trust that those with specialised experience which has taken a lifetime to acquire will excuse what they may see as my naivety of treatment of their subject. I felt them looking over my shoulder frequently as I laboured at the work. Those who helped were as follows.

Dr James Bathurst (Newcastle University) commented on my summary of the slope-area method of estimating river flow. *John Bell* (ex IH, retired) supplied notes describing soil moisture measurement by the thermogravimetric method. *Ken Blyth* (IH) gave up-to-date advice on remote sensing satellite hardware. *Prof. Chris Collier* (Salford University) read the section on weather radar and suggested changes and additions. *Dr J. David Cooper* (IH) discussed the capacitance probe and time domain reflectometry for measuring soil moisture and gave access to instruments to produce Figs. 9.5(a), (b), 9.6 and 9.8. *Andy Dixon* (IH) patiently guided me through the complexities, and terms, of borehole drilling, and loaned me two photographs (Figs. 9.15 and 9.16). *Eumetsat* supplied up-to-date information on Meteosat services. *Dr John Gash*

(IH) advised on the latest techniques in evaporation (and CO_2) flux measurement, gave access to the equipment shown in Figs. 5.10, 6.2, 6.3 and 6.4 and checked the chapter on evaporation. *Dr Reg Herschy* (CNS, Reading) read the chapter on water and suggested changes, in particular concerning river flow measurement. *Wynn Jones* (Met. Office) supplied much information on ocean buoys, provided Figs. 13.2, 13.3 and 13.4 and checked the section on ocean measurements. *Robin Pascal* (Southampton Oceanography Centre) spent an afternoon answering questions regarding instrumentation for oceanographic research and supplied papers on this. *Dr Richard Pettifer* (Vaisala UK Ltd) made comments on the latest state-of-the-art radiosonde sensors and their wind measurement techniques. *Dr Jonathan Shanklin* (British Antarctic Survey), who has been to Antarctica many times, commented on the meteorological instrumentation currently in use on that continent. *The Royal Meteorological Society* gave access to instruments, resulting in the photographs for Figs. 7.2(*a*),(*b*), 7.3(*a*),(*b*) and 7.4. *Dr John Stewart* (ex IH, now at Southampton University) read, and suggested changes to, the chapter on remote sensing. Finally, my copy-editor, *Dr Susan Parkinson*, suggested numerous improvements. The book is that much better for their help.

1
Basics

The need for measurements

Whether it be for meteorological, hydrological, oceanographic or climatological studies or for any other activity relating to the natural environment, measurements are vital. A knowledge of what has happened in the past and of the present situation, and an understanding of the processes involved, can only be arrived at if measurements are made. Such knowledge is also a prerequisite of any attempt to predict what might happen in the future and subsequently to check whether the predictions are correct. Without data, none of these activities is possible. Measurements are the cornerstone of them all. This book is an investigation into how the natural world is measured.

The things that need to be measured are best described as *variables*. Sometimes the word 'parameters' is used but 'variables' describes them more succinctly. The most commonly measured variables of the natural environment include the following: solar and terrestrial radiation, air and ground temperature, humidity, evaporation and transpiration, wind speed and direction, rainfall, snowfall, and snow depth, barometric pressure, soil moisture and soil tension, groundwater, river level and flow, water quality (pH, conductivity, turbidity, dissolved oxygen, biochemical oxygen demand, the concentration of specific ions such as nitrates and metals), sea level, sea-surface temperature, ocean currents and waves and the ice of polar regions.

The origins of data

Early instrument development

Measurement of the natural environment did not begin in the scientific sense until around the middle of the seventeenth century; in 1643, working in Florence, Torricelli made the first mercury barometer, based on notes left by

Galileo at his death. The first thermometer is also attributable to Galileo. Castelli, also in Italy, made the first measurements of rainfall. Sir Christopher Wren in England designed what was probably the first automatic weather station, while Robert Hooke constructed a manual raingauge. In 1846 Thomas Robinson, a clergyman in Armagh, constructed the first instrument for measuring wind speed and in 1853 John Campbell developed the prototype of today's sunshine recorder. Thirteen years later, Thomas Stevenson, the father of Robert Louis Stevenson, designed the now widely used wooden temperature screen. In 1850 George Symons embarked on a lifelong mission to put rainfall measurement on a firm footing, while Captain Robert FitzRoy, who had earlier taken Charles Darwin on the voyage of the *Beagle*, spent much of his later career advancing meteorological measurements at sea. Thus the beginnings of the study of the natural environment started with the development of instruments and the taking of measurements, highlighting the great importance of hard facts in forwarding any science and of the means of obtaining these data. Instrument development continued into and through the twentieth century using similar technology until, mid-century, the invention of the transistor presented totally new possibilities.

The important point about the early designs is not their history, interesting as it is, but that the same designs, albeit perhaps refined, are still in widespread use today. Most of the national weather services (NWSs) of the world rely on them. All the data used by climatologists to study past conditions, and by anyone else for whatever purpose, are derived largely from instruments developed in the Victorian era, and the same instruments look set to continue in widespread use into the foreseeable future. The importance of appreciating their capabilities and limitations is thus of more than passing historical interest. Everyone using data from the past and from the present needs to be aware of where the data come from and of how reliable or unreliable they are likely to be.

Recent advances

In the last thirty years, owing to developments in microelectronics and the widespread availability of personal computers (PCs), new instruments have become available that greatly enhance our ability to measure the natural environment. It has been possible to design data loggers with low power consumption and large memory capacity that can operate remotely and unattended, and new sensors have been developed to supply the logging systems with precise measurements. (Sensors may be referred to in some texts as transducers.) In a balanced review of the subject, this new generation of instruments needs to be discussed along with the old, for we are at a time in the measurement of the environment when both types of instrument are equally important, although the change to the new is well under way and accelerating.

Old and new compared

A serious limitation of the old instruments is that, being manual and mechanical, they need operators and this restricts their use to those parts of the world that are inhabited. Most mountainous, desert, polar and forested areas and most of the oceans are, in consequence, almost completely blank on the data map. Thus far, our knowledge of the natural environment comes from a rather limited range of the planet's surface.

In contrast the new instruments need the attendance of an operator only once every few months, or even less frequently, and so can be deployed at remoter sites. The same electronic developments have also made it possible to telemeter measurements from remote automatic stations via satellites, making it possible to operate instruments in almost any region of the world, however remote. To this capability has also been added the new technique of remote sensing – its images being generated by the same satellites as those that relay data.

The old instruments also lack the accuracy of the new. Compare for instance the data from a sunshine recorder giving simple 'sun in or out' information with those from a photodiode giving an exact measure of the intensity of solar energy second by second. Ironically, these improvements are often a hindrance to change, for although changing to a better instrument improves the data the continuity of old records is lost. This deters many long-established organisations from changing their methods and so the old ways persist. While it is possible to operate a new instrument alongside the old to establish a relationship between them, this is usually only partly successful, the complexity of the natural environment and of an instrument's behaviour meaning that a simple relationship between the two rarely exists.

The new instruments record their measurements directly in computer-compatible form, in solid state memory, allowing their measurements to be transferred to a portable PC or retrieved by removing a memory unit. In the case of telemetry the transmission, reception, processing and storage of the measurements is fully automatic. There is none of the labour-intensive work involved with handwritten records or with reading manually values from paper charts.

Because automatic instruments do not require the regular attendance of an operator and because their data can be processed with the minimum of manual intervention, staff costs are also reduced.

Both old and new instruments are thus important at this stage in the evolution of environmental monitoring and both are covered in the chapters that follow.

General points concerning all instruments

Definitions of terms

For any instrument, the *range* of values over which it will give measurements must be specified. For example a sensor may measure river level from 0 to 5 metres. (In old terms, 5 metres would have been called the instrument's full-scale deflection, meaning that the pointer on an analogue meter has been deflected to the end of its scale.)

The *span* is the difference between the upper and lower limits. If the lower end is zero, the span is the same as the range, but if the range of a thermometer is, say, $-10°C$ to $+40°C$ its span is $50°C$.

The *accuracy* is usually expressed in the form of limits of uncertainty, for example a particular instrument might measure temperature to within $±0.3°C$ of the actual temperature. Accuracy may differ across the range covered, such that in the case of a relative humidity (RH) sensor, the accuracy may be $±2\%$ RH from 0–80% RH while from 80%–100% RH it may be $±3\%$.

The *error* is the difference between an instrument's reading and the true value. It is thus another way of stating accuracy. A systematic error is a permanent bias of the reading in one direction, away from the correct value, as opposed to a random error, which may be plus or minus and might, therefore, cancel out over a series of readings.

Instruments often respond to more than just the variable they are meant to measure. Temperature dependence is, for example, a common problem. In such a case a *temperature coefficient* will be quoted to allow a correction to be made. The coefficient is given as the percentage change in reading per degree Celsius change of temperature. Not only may the sensitivity change with temperature, but so also may the zero point.

The *resolution* of an instrument indicates the smallest change to which it responds; for example, a meter might respond to river level changes of as little as 1 mm. This is also sometimes known as sensitivity or discrimination. But just because an instrument can respond to changes of 1 mm, it does not necessarily mean that it measures them to that accuracy; the instrument might only have an accuracy of $±2.0$ mm.

Sometimes an instrument does not respond equally across its full working range, so that its response deviates from a straight line. The extent of the deviation is expressed as its *linearity*, this usually being specified as the maximum deviation (as a percentage of the full scale reading) of any point from a best-fit straight line through the calibration points. The term is used only for instruments that have a close-to-linear response. Some sensors have characteristics that are far from approaching a straight line and may not even have a smooth-curve response. In the case of mechanical instruments, non-linearity may be compensated for in the design of the mechanical links between the sensor and the recording pen; in the case of electrical instruments, a non-linear response may either be compensated electronically, or by recourse to a look-up table or an equation when processing the records in a computer.

Hysteresis is the characteristic of a sensor whereby it responds differently to an increasing reading and to a falling one, the upward curve following a different path to that downward. This might be due to 'stiction' in the mechanical links between sensor and pen; there are equivalent processes in electronic systems. Again the deviation is expressed as the percentage difference (of full scale) between increasing and falling readings.

With the passage of time, a change in instrument performance often occurs, and settings *drift*. This may be due to the aging of components, which results in a change in sensitivity or a change in the zero point, or both. Stability is the converse of drift and is usually expressed as the change which occurs in a sensor's sensitivity, or zero point, over a year. Because of drift, it is usually necessary to readjust, or *calibrate*, an instrument periodically.

Sensors take time to respond to a change in the variable which they are measuring. Some respond in milliseconds, others take minutes. A *time constant* is given to quantify this. This is the time for a reading to increase to $1 - 1/e$ (about 63%) of its final value, or to fall to $1/e$ (about 37%) of its initial value, following a step change in the variable (Fig. 1.1). It takes about three to four times longer than the time constant to reach 95% of the final reading. Thus a constantly changing variable with a slow-response sensor can result in considerable error. To complicate matters further, the response in one direction can be different to the reverse, owing to hysteresis, as mentioned above. These various terms are illustrated in Fig. 1.2.

The *repeatability* is the degree of agreement between an instrument's readings when presented more than once with the same input signal. Different readings may result each time through a combination of all the various sources of error.

There is also the question of the number of decimal points shown. As mentioned above, just because, for example, a temperature is quoted as being 25.03°C it does not necessarily mean that it is accurate to hundredths of a degree. The instrument may only be good to ± 0.1 degrees or less. Extra

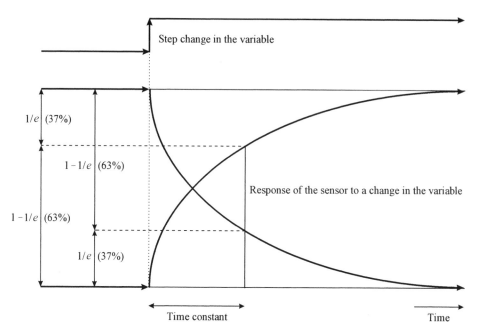

Figure 1.1. The meaning of time constant for response of a sensor to a variable.

decimal points can be introduced into data in many ways, for example during processing to fit a standard format. Decimal points can also be lost in the same way. Users of data should always beware of such possibilities.

Choice of site

Where, and how, an instrument is placed in the field (finding a representative site and positioning the instrument on the site) affects how good its readings are, often having as great an influence on overall accuracy as the instrument's own characteristics. Comments are made on this throughout the book. Changes at a site after an instrument has been installed, perhaps slowly over years such as the growth of trees or suddenly such as the erection of buildings, also need consideration.

Maintenance

The maintenance of an instrument, such as its periodic readjustment against a standard, the regular painting of a temperature screen or the levelling of a raingauge, is as important as any other aspect of accuracy and data quality. Neglect of this is widespread and results in unreliable data.

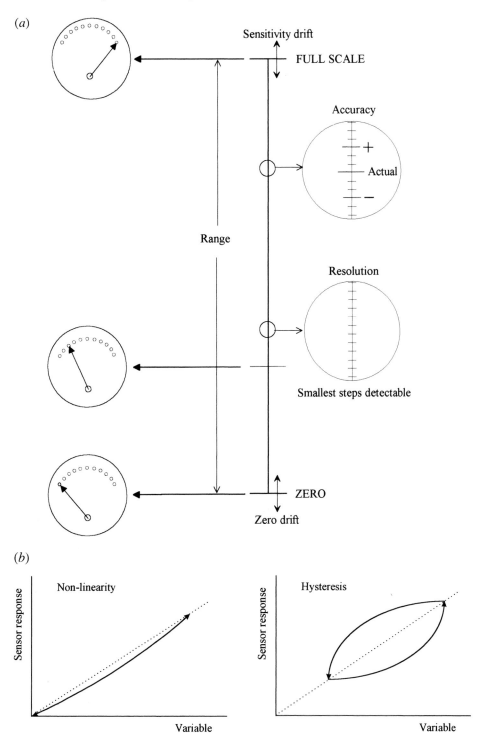

Figure 1.2. Illustration of terms.

Spatial variability

Most environmental variables vary not just with time but also across space, for example across a local thunderstorm. Faced with this problem, it can be tempting to conclude that it is pointless to improve instrument performance because instrument error is swamped by spatial variability as well as by the difficulty of selecting the perfect site. But this is a philosophy of defeat. Better, instead, to improve the instrument, choose the site more carefully, take measurements at several points and so get an estimate of variability. Spatial variability cannot be measured unless we have good instruments, for what might be assumed to be spatial variability may instead be instrument error.

Using data of unknown quality

It is difficult to know how good data are if there is no record, along with the data, of what type of sensor was used, how it was exposed or its maintenance schedule (metadata). Often there is no sure way of checking on these matters. In general, our data sets are a mix of measurements collected by a variety of instruments, installed in different ways and operated at different levels of competence, all of which may have changed with time and with the operator. Discontinuities and different levels of error are invariably present.

Plan of the book

The aim of this book is to show how measurements of the natural environment are made and to point out the pitfalls. Each variable, or group of variables, is given a chapter to itself, working through from the early instruments, most of which are still in use, to the latest developments, only a few of which may yet be in use. How the measurements from the new electronic sensors are processed and stored automatically in data loggers or are telemetered to a distant base are the topics of two later chapters. While remote sensing is a specialised topic, already well covered in other publications, a chapter on it is included for completeness, comprising how measurements are made remotely, how good or bad they are and how they relate to measurements made *in situ* on the ground.

Mathematical analysis is used sparingly, and where equations are used a line-by-line derivation from first principles is not always given. The SI (Système International d'Unites, NPL 1963, BSI 1969) is used throughout, unless otherwise stated. The method of defining units is, for example, to abbreviate metres per second to $m\,s^{-1}$ and cubic metres to m^3.

This is not an operator's handbook, for the technician, but an outline for the users of data who would like (or need) to know how the data they use were collected and how good or bad they are likely to be. It is also intended for instrument designers who need background information, or for those about to collect field data and instal instruments, or for students of meteorology, hydrology and environmental sciences, who should know how measurements are made. Useful additional information on the basic meteorological instruments can be found in three UK Meteorological Office handbooks (1982a, b and 1995) while the World Meteorological Organisation (WMO 1994) periodically issues and updates technical reports on how meteorological instruments should be installed and operated.

References

British Standards Institute (1969) The use of SI units. Report PD 5686. HMSO, London.

Meteorological Office (1982a) The handbook of meteorological instruments. HMSO, London.

Met. Office (1982b) The observer's handbook. HMSO, London.

Met. Office (1995) The marine observer's handbook. HMSO, London.

National Physical Laboratory (1963) The inclusion of metric values in scientific papers. HMSO, London.

World Meteorological Organisation (1994) Guide to meteorological instruments and methods of observation. WMO No. 8 (generally known as the CIMO Guide).

2
Radiation

The variable

Integrated over the whole of its radiation spectrum, the sun emits about 74 million watts of electromagnetic energy per square metre. At the mean distance of the earth from the sun, the energy received from the sun at the outer limits of the earth's atmosphere, at right angles to the solar beam, is about 1353 watts per square metre ($W\,m^{-2}$) and is known as the solar constant. In fact the energy received is not quite constant but varies over the year by about 3%, because the earth is in an orbit around the sun that is actually elliptical. The actual output of the sun itself also varies with time, the most familiar regular rhythm being the 11-year sunspot cycle, although the variations due to this are less than 0.1%. There are other, longer, cycles such as the 22-year double sunspot cycle, and the 80–90-year cycles (Burroughs 1994). It is useful to define some terms.

Units and terms

Radiant flux is the amount of electromagnetic energy emitted or received in unit time, usually expressed in watts (1 watt = 1 joule per second).

Radiant flux density is the radiant flux per unit area expressed in watts per square metre ($W\,m^{-2}$), although other units such as $mW\,cm^{-2}$ and $cal\ cm^{-2}\,min^{-1}$ may be used ($1\,W\,m^{-2} = 0.1\,mW\,cm^{-2} = 10^{-3}\,kW\ m^{-2} = 1.433 \times 10^{-3}\,cal\,cm^{-2}\,min^{-1}$). No longer in wide use is a unit called the langley, which is equivalent to $1\,cal\,cm^{-2}$.

Irradiance is the radiant flux density received by a surface; the *emittance* is the flux density radiated.

Insolation is the irradiance integrated over a period of time, typically a day in the case of solar radiation; thus the insolation is the total energy received in the period. On cloudless days the diurnal variation of solar irradiation is roughly sinusoidal. In the UK in summer the maximum intensity of the radiation at solar noon is about 900 W m^{-2}. With a day length of 16 hours, the maximum daily insolation can be calculated to be about 9 kWh m^{-2}. However, insolation is often expressed in megajoules rather than kilowatt hours, 1 kWh being equivalent to $1000 \times 60 \times 60 = 3.6$ MJ, giving a UK daily maximum insolation of about 32 MJ m^{-2}. This is reduced by cloud cover to a summer average of 15 to 25 MJ m^{-2}, while in winter it varies from 1 to 5 MJ m^{-2}.

Spectral composition

Figure 2.1 illustrates the spectrum of the sun's emission, both as received just outside the atmosphere and as received at the earth's surface after modification by its passage through the atmosphere (see also Fig. 14.1). During its passage, radiation in different parts of the spectrum is absorbed, scattered and reflected. For example, some ultraviolet radiation is absorbed by ozone and by oxygen while parts of the infrared spectrum are absorbed by water vapour and carbon dioxide. Scattering and diffuse reflection occurs from particles larger than the wavelength of the radiation, such as dust, smoke and aerosols from volcanic activity, while Rayleigh scattering by air molecules and particles smaller than the wavelength gives rise to the blue sky and the blue appearance of the earth from space. Clouds can reflect up to 70% of the energy from their upper surface.

As a consequence of scattering and reflection, the irradiation received at ground level is in the form of both direct and diffuse radiation. In cloudless conditions, the ratio of the diffuse to direct radiation depends on the angle of the sun and on the amount of aerosols in the atmosphere, the diffuse component varying from 10% in very clear air to 25% in polluted air; typically it is about 15% (Monteith 1975). Just before sunrise, just after sunset and in cloudy conditions, the radiation is of course entirely diffuse.

To differentiate between the radiation from different sources, various categories of radiation have been defined (Meteorological Branch, Dept of Transport, Canada 1962). The wavelength limits usually quoted are somewhat arbitrary, particularly at the longer infrared wavelengths, and should be seen only as a guide and not as a precise statement.

Solar radiation is the radiation received from the sun reaching the measuring instrument without a change in wavelength, sometimes also referred to as short-wave radiation (Fig. 2.1). At the surface of the earth, most of the radiation is in the spectral band from 0.3 to 3.0 µm.

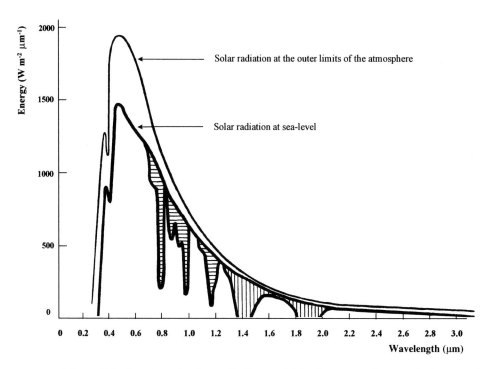

Figure 2.1. Spectra of solar radiation outside the atmosphere and at ground
level showing the bands where there is absorption due to water vapour and
carbon dioxide in the mid-IR (vertical shading) and due to oxygen and water
vapour in the mid- and near-IR (horizontal shading), while oxygen and ozone
absorb in the UV at wavelengths shorter than 0.4 μm (not shaded).

Terrestrial radiation is that emitted by the ground and also by the atmosphere, in particular by water vapour and carbon dioxide – the same gases that
also absorb it – reaching the instrument without change of wavelength within
the (abitrary) limits of 3 to 100 μm; it is sometimes referred to as long-wave
radiation. Included in this is radiation emitted by clouds in the narrow band
from 8 to 13 μm, a region in which little radiation is emitted by water vapour
and carbon dioxide. These two general categories are further subdivided as
follows.

Direct solar radiation refers to the incoming direct solar energy incident on a
surface at right angles to the solar beam. (All other definitions refer to the
radiation falling on a horizontal surface.)

Diffuse solar radiation is that part of the incoming solar energy incident on a
horizontal surface shielded from direct radiation from the sun.

Hemispheric solar radiation (also known as *global radiation*) is the sum of
both the direct and the diffuse radiation on a horizontal plane.

Reflected solar radiation is that portion of the total solar energy reflected from the ground (and atmosphere) upwards without a change of wavelength, measured on a downward-facing horizontal surface. The ratio of this to the total incoming solar radiation is the *albedo* of the surface, more correctly known as the *reflection coefficient* of the ground (which must be differentiated from the surface's *reflectivity*, the fraction of the total incoming solar radiation reflected at a specific wavelength).

Net radiation is the difference between all incoming and all outgoing radiation fluxes of all wavelengths (solar and terrestrial) on a horizontal surface.

Total incoming radiation is the sum of all incoming solar and terrestrial radiation.

Total outgoing radiation is the sum of all outgoing solar and terrestrial radiation.

Incoming long-wave radiation is that radiated downwards from the atmosphere and clouds.

Outgoing long-wave radiation is that radiated upwards from the ground and atmosphere.

Photosynthetically active radiation (PAR), covering the band from 0.4 to 0.7 μm, is the spectrum used by plants for photosynthesis.

Ultraviolet (UV) radiation extends from about 0.1 to 0.4 μm and is subdivided into three narrower bands designated A, B and C.

Daylight illumination refers to the visual quality of the radiation in terms of the spectrum to which the human eye responds, which is from 0.38 to 0.78 μm peaking at 0.55 μm.

Sunshine recorders

The Campbell–Stokes recorder

Since the maximum possible level of irradiance at noon for any one place and for any particular date is known, and since the daily variation is approximately sinusoidal, a knowledge of cloudiness will give some indicatgion of daily insolation. A sunshine recorder gives information not only on sunshine duration but also on the extent of any cloud, from which a rough estimate of daily insolation can, therefore, be derived, and which may be relevant to climatic change.

The first attempt to record sunshine duration was made in 1853 by John Campbell, who used a glass sphere to burn a trace of the sun's image focused onto the inside of a wooden bowl. This was modified in 1880 by Sir George

Figure 2.2. The sun's image focused onto the card of a Campbell–Stokes sunshine recorder burns a track as the sun moves across the sky.

Stokes, hence its present-day name, the Campbell–Stokes recorder. The modifications by Stokes, and subsequent refinements made by manufacturers, led to the present-day design (Fig. 2.2) in which the sun's image is focused onto a specially manufactured card by a carefully ground glass sphere, which burns a track along the card if the sun is not obscured by cloud (Fig. 2.3). The card is manufactured so that it does not catch fire, but simply chars, and is affected minimally by rain. This simple instrument is still in widespread use worldwide.

If there is uninterrupted sunshine, or none, the trace is easy to read. Rules have been established for interpreting the meaning of the burnt track but the estimate can only be approximate when there are intermittent, short bursts of sunshine or when the sun is hazy. There is also some uncertainty around sunrise and sunset. Further error is introduced through imprecise adjustment of the instrument, notably with regard to the exact positioning of the sphere relative to the card.

For use at different latitudes and in different seasons, a selection of card holders and cards of varying shape and size is needed (Fig. 2.3), there being models covering latitudes 0° to 45° and 45° to 65° (N or S). For latitudes higher than 65°, two recorders must be operated back-to-back, one being extended in latitude by simple modification and the other being both modified and set on an inclined wedge. However, there is a tendency today to use electrical sunshine sensors rather than this complicated dual arrangement.

It is difficult to put a figure on the accuracy of this type of sensor, but it is unlikely to be better than $\pm 5\%$ since the interpretation of the track is to some extent subjective.

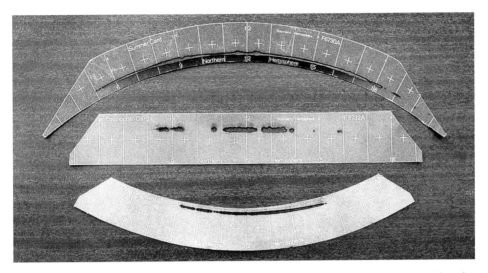

Figure 2.3. Traces burnt on summer, winter and equinoctial cards of a Campbell–Stokes recorder. Upper, 1 May 1997; centre, 4 October 1997; lower, 1 January 1998 (Institute of Hydrology site in southern England).

Electrical sunshine sensors

Because of the long records of sunshine made by the Campbell–Stokes design, going back to the nineteenth century, there was a need in some quarters to continue sunshine observations into the future, but on an automatic basis. To satisfy this requirement, electrical sunshine sensors were developed.

In the Haenni type, a narrow shutter rotates, several times per second, around a circular array of light detectors, shutting off the direct solar beam momentarily in turn to each detector that has a view of the sun. If the sun is not visible owing to cloud, the passage of the shutter causes little change to the light level detected since the shutter obscures only a small part of the diffuse radiation. If the sun is visible, however, a large change occurs momentarily as the sun becomes shadowed by the shutter. In another, similar, design a slit is used in place of a shutter. Thresholds have to be set to enable the sensor to detect the transition from 'no sun' through 'hazy sun' to 'full sun', and some uncertainty will inevitably be introduced. However, the instrument responds instantaneously to the presence or absence of the direct solar beam, which is an improvement on the Campbell–Stokes instrument, but it requires power to drive the shutter motor and this makes it less suited to remote sites where power is usually limited. To prevent ice formation, the glass cover can be heated, but this also consumes power. If the Campbell–Stokes instrument is frost- or snow-covered it will not operate, but there is an operator to clear it; however, the operator will not be there all day, every day.

The Shiko type of sensor uses two photoresistor cells exposed to the sun in a diffusing glass tube, one being shielded from the narrow band of sky across which the sun passes. In the absence of direct sun the two sensors are of very similar resistance, but they are considerably different when the sun is visible. This unbalance is detected and gives a yes/no signal that can be recorded. It requires no power and responds rapidly (Amano *et al.* 1972).

Meant mostly for amateurs, the Jordan recorder consists of two semicylinders with slits on their flat side, one facing east and the other west, their angle adjustable for latitude. The presence or absence of direct sunlight is recorded on light-sensitive paper of the blueprint type. The records are said to be open to more uncertainty than the Campbell–Stokes recorder because of the difficulty of ensuring that the sensitivity of the paper is consistent, but one would have thought that this could be overcome without much difficulty.

Other types exist, such as the 'Marvin' design, but they are not true sunshine sensors because they cannot differentiate well enough between diffuse and direct radiation.

Compatibility between sunshine recorders

Any change from one type of sunshine recorder to another inevitably results in a discontinuity of readings (as it does with any instrument). Differences of up to 20% can occur if a change is made from the Campbell–Stokes to any of the electrical sensors (World Meteorological Organisation 1994, Chapter 1 references). The Campbell–Stokes instrument is the classical and most used sunshine recorder and was selected by the WMO in 1962 as the reference standard against which other types should be compared. By reduction of sunshine totals to this standard the WMO believes that systematic differences can be reduced to ±5%.

Radiometers

'Radiometer' is the generic term for all instruments measuring radiation. In the case of solar and terrestrial radiation, they are subdivided into several classes depending on the part of the spectrum they sense, as follows.

Pyranometers measure all incoming solar radiation, direct and diffuse, on a horizontal surface within a field of view of 180°. They are also known as *solarimeters*.

Effective pyranometers measure the same thing as pyranometers except that they include terrestrial long-wave radiation as well as solar. For this reason they are sometimes called *total radiometers*.

Pyrheliometers measure just the direct solar radiation, and do so at right angles to the solar beam, the instrument tracking the sun's movement. The sensing element is shielded so that it sees only the sun and a small annulus of sky around it.

Net pyrradiometers sense the difference between the total incoming and total outgoing radiation, of both solar and terrestrial origin, that is the net value of the upward and downward fluxes of both long- and short-wave radiation. They are also known as net radiometers and radiation balance meters.

Pyrgeometers respond to long-wave terrestrial radiation only. They can be upward- or downward-looking, depending on whether incoming or outgoing radiation is to be measured. They are also known as infrared radiometers.

Photometers measure that part of the spectrum to which the human eye is sensitive, 0.38 to 0.78 µm.

Photosynthetically active radiation (PAR) sensors measure the spectral band used by plants to photosynthesise, 0.4 to 0.7 µm.

Ultraviolet sensors measure from 0.2 to 0.4 µm. They do not have a special name.

The bimetallic actinograph

An early step towards measuring solar energy, rather than sunshine duration, was the development of the *bimetallic actinograph*, also known as a *pyranograph* because it is a pyranometer which records on a chart. (The term 'actinograph' comes from the word 'actinic'; the property of *actinic* rays, especially the shorter wavelengths of green to ultraviolet, is to produce photochemical effects, although their measurement need not be based on chemical changes.)

In an actinograph, a blackened bimetallic strip exposed horizontally to the sun beneath a protecting dome bends in proportion to the solar energy incident on it. A bimetallic strip is made by welding together two bars of dissimilar metals and rolling them into a thin strip. When the temperature changes, the two strips expand by different amounts and the composite strip bends; (see also the *bimetallic thermograph*, Chapter 3). One end of the strip is fixed and the other can move and as it bends the motion is transferred to a pen via levers, recording a trace on a paper chart fixed to a drum that is rotated, usually once a day, by a spring- or battery-operated clock. To prevent air-temperature changes from affecting the record, the fixed end of the strip is supportedly by a second bimetallic strip, shielded from the sun within the case, which responds only to changes in ambient temperature. In addition a third, shorter, bimetallic strip, at right angles, ensures that the sensing strip remains horizontal so that its angle to the sun does not change. In an alternative design, temperature compensation is achieved by two white bimetallic strips exposed on either side of the black strip, which reflect most of the solar radiation and so bend mainly in proportion to air temperature changes (Latimer 1962).

This type of instrument takes up to five minutes to respond to changes in radiation, in part due to pen 'stiction' but also because of the time taken by the strip to heat or cool. It is not, therefore, a precise instrument, nor can it be used to measure instantaneous levels of solar radiation. However, by measuring the area beneath the curve recorded on the chart throughout the day (with a *planimeter*) a rough estimate of the daily total of energy received can be obtained – to about $\pm 10\%$. Actinographs are now little used, because of the availability of much superior, and as cheap, electrical methods.

Pyranometers

Thermal solarimeters   There are two kinds of solarimeter – thermal and photoelectric. The thermal types, like their mechanical counterparts, sense radiation intensity by measuring the heating effect of the radiation, but they do it electrically. All are similar in principle, having a black surface exposed within a protecting dome, the black surface rising in temperature by an amount proportional to the direct and diffuse radiation it absorbs. By correct choice of black material, absorption can be made equal across the full solar spectrum. The black surface is exposed beneath a glass dome, which, being transparent to all but the far ultraviolet and opaque to any incoming infrared radiation of terrestrial origin, allows just the solar spectral band to reach the sensing element.

The measurement of the rise in temperature of the black surface is relative to the ambient air temperature and this is achieved by measuring the temperature difference either between the black surface and the sensor case, shielded from the radiation (Fig. 2.4), or between the black surface and a similar, but white, surface alongside it (Fig. 2.5). The black surface has traditionally been a coating of lacquer containing carbon black, which produces a matt black finish with wide spectral absorption. Magnesium oxide has been used to produce the white surface because it reflects the solar spectrum very effectively while absorbing wavelengths longer than $3\,\mu m$. The latter is important because, although not gaining access from the outside, long-wave infrared radiation does originate within the glass dome itself (since all bodies above absolute zero radiate energy in a frequency band dependent on temperature). By using magnesium oxide as the cold-junction coating, the long-wave radiation originating from within the glass dome is absorbed by both the black and white

Figure 2.4. The black (hot) sensing surface beneath the double glass domes of a thermal solarimeter of the shielded type; a white conical cover keeps the case at ambient air temperature (cold junction).

Figure 2.5. The alternate black (hot) and white (cold) sensing surfaces of a 'star' type of solarimeter beneath its single glass dome.

surfaces alike, thereby (largely) cancelling out. The effect of long-wave radiation that emanates from within the glass dome is minimised in designs which have only a black surface (Fig. 2.4) by using two concentric domes; the inner one reduces convective and radiative heat exchange between the sensor and the outer dome. In some designs the outer dome can be interchanged, allowing materials with different spectral-transmission characteristics to be substituted for the detection of specific bands of radiation. Some designs also ventilate the

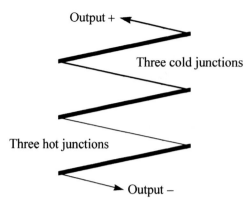

Output +

Three cold junctions

Three hot junctions

Output −

Figure 2.6. A thermopile is a series of junctions between two dissimilar conductors. If one set is maintained at one temperature and the other at a different temperature, a (very small) voltage is generated that is proportional to the difference in temperature. This is useful where differences in temperature (rather than absolute temperature) are to be measured, radiation sensors being a good example.

outer dome by circulating warm air over it to prevent the formation of frost and dew or the accumulation of snow. The resultant extra heating and consequent long-wave radiation is taken care of through the use of a double dome or the use of magnesium oxide as the white surface.

In both solarimeter designs it is the temperature difference which is sensed and this is most conveniently measured with a thermopile (Fig. 2.6). In both designs, the hot junctions of the thermopile are in contact with the black surface, while the cold junctions either make thermal contact with the shielded body of the instrument or with the white surface exposed alongside the black (Latimer 1962). The thermopiles are typically manganin–constantan, antimony–bismuth or copper–constantan, the number of junctions ranging from 10 to 50. Today, thick-film microelectronic techniques can simplify the production of thermopiles. The resultant outputs from such sensors are in the low-millivolt range and need amplification before logging, recording or displaying. Some recent designs use a circular array of diodes in a bridge circuit to measure the temperature difference in place of the thermopiles and these can produce outputs 20 or more times greater.

As the sun moves across the sky, the angle of incidence of the direct beam varies. The sensor response should be proportional to the cosine of the angle, being zero when the sun is on the horizon. However, owing to variations in the black surface, particularly at low sun angles, and also to refraction through the glass domes, cosine errors of about $\pm 2\%$ can occur. Furthermore, as the azimuth of the sun (the rotational angle measured eastwards from the north–south direction) varies, the sensitivity of the sensor may vary, although the error is reduced if the sensor is rotationally symmetrical. A typical time constant for this type of sensor is 5–20 seconds. If all errors are considered together, an overall accuracy of about ±5% is achievable, but only if care is taken in calibrating, installing and operating the instrument.

Two modifications can be made to solarimeters to extend the range of measurement. A *shadow ring* can be fitted to cast a shadow of the sun on the sensor, thereby giving a measure of just the diffuse radiation. To measure the albedo of the ground, two sensors are used, one facing up and the other down. If their outputs are recorded separately, a subsequent calculation will give the ratio of the two readings and thus the albedo (or the reflection coefficient). If they are connected in opposite polarity and the single composite output is recorded, a measure of the short-wave radiation balance is obtained.

Pyranometers can also be used as pseudo-sunshine sensors, by taking readings every minute and comparing them with a threshold (a value agreed internationally as 120 W m^{-2}) to differentiate between sunshine as opposed to daylight. However, there is some uncertainty as to the correlation between the readings produced by this method and those from a Campbell–Stokes recorder.

By suitable data processing, day length can also be sensed, again by setting a threshold against which the level of irradiance can be compared. These additional functions can be achieved at the same time as using the sensor as a pyranometer.

Photodiode sensors   The light-sensitive diode, photodiode or photovoltaic cell is a useful new sensor capable of measuring solar radiation (Fig. 2.7). It functions quite differently from thermal sensors, responding to the individual incoming photons directly and producing free charge carriers – electrons and holes (a 'hole' is the absence of an electron, effectively a positive charge). It can be in the form of a p–n junction depletion-layer diode or can have a third layer sandwiched between the p and n layers. There are also Schottky photodiodes, which use a metal–semiconductor junction. The open-circuit voltage of photodiode sensors rises rapidly with increasing radiation levels, soon reaching saturation at around 0.6 volts, but their short-circuit current is proportional to the irradiation level over a wide range of intensities and it is this characteristic that is used to measure solar radiation. Because this type of sensor acts at the atomic level, producing an electrical signal directly, its time constant is measured in microseconds.

Such diodes have the advantage of being cheaper than sensors of the thermal type. However, they have the drawback that their spectral response is not flat, nor does it encompass the full solar spectrum, ranging only from 0.3 to 1.1 μm (Fig. 2.8). Some photodiode solar sensors are blue-enhanced by

Figure 2.7. The light-sensitive diode solarimeter. Beneath the flat, white diffusing cover is a light-sensitive diode, offering a cheaper alternative to the thermal type of solarimeter.

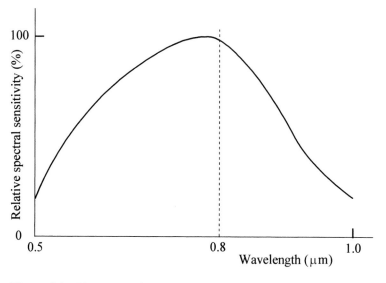

Figure 2.8. The spectral response of a photodiode, which is narrower than that of thermal radiation sensors, in which the black surfaces absorb virtually all the incoming radiation irrespective of wavelength. In the case of a basic silicon photodiode, as shown here, the sensitivity ranges from the violet into the infrared, peaking at around 0.8 μm in the near-infrared. Its performance as a solarimeter will thus depend partly on the spectral composition of the radiation at any given time.

filters to compensate for the reduced sensitivity at these shorter wavelengths (Chappell 1976). The error associated with the limited spectral range is from $\pm 3\%$ to $\pm 5\%$, the actual value depending on the spectral composition of the radiation, which varies with season, time of day and cloud cover. Photodiodes have good long-term stability but a temperature coefficient of around $0.1\%\,^{\circ}C^{-1}$, for which compensation must be made. Their overall accuracy is somewhat less than that of a well-serviced thermopile pyranometer, but their much lower cost often outweighs this consideration.

PAR sensors

The spectral range of light-sensitive diodes is sufficient to cover all that part of the spectrum associated with photosynthesis, that is from 0.4 to 0.7 μm. With suitable filtering to mimic the response of plants, diodes can be used very effectively as sensors of *photosynthetically active radiation* (PAR). Diodes are also well suited to this task since they measure the incoming quanta of energy directly and this is closely related to the rate of sugar production in plants. The units of measurement used are not watts per unit area but moles per unit area; a mole (abbreviated in units to 'mol') is a measure of the amount of a substance, in this case of photons, and in this application it is expressed in the form $\mu mol\ m^{-2}\ s^{-1}$. (See also pH, Chapter 10.)

Photometers

The spectral range of photodiodes also covers that of the human eye (0.38 to 0.78 μm) and with suitable filters can be made to simulate the eye's response so as to measure natural (daylight) illumination. The unit of illumination is the lux $= 1$ lumen m^{-2}. 1 lumen $= 0.001496$ watts, therefore 1 lux $= 0.001496$ W m^{-2} (or 668.5 lux $= 1$ W m^{-2}). 1 foot-candle $= 10.76$ lux.

Ultraviolet sensors

Heterojunction photodiodes, which are based on two different semiconductors having different band gaps, can measure very specific radiation bands. If the materials are, for example, gallium arsenide and germanium the ultraviolet (UV) spectrum can be measured. (The simpler diodes described earlier do not respond to these shorter wavelengths.) Such UV diodes are made covering three bands. The C-band peaks at 0.29 μm; this is the band that causes most erythema (sunburn) and other biomedical processes such as vitamin D production. This is hard UV, and is much influenced by stratospheric ozone. The B-band peaks at 0.31 μm, but is still noticeably affected by the ozone layer. The

A-band peaks at 0.34 μm and is little affected by ozone, behaving more like blue light but with increased scattering. The use of the three sensors together can give information about the effects of cloud and aerosols on the trans- mission and scattering of UV radiation across the full UV spectrum.

UV sensors are similar in design and appearance to sensors using normal photodiodes; the protective covers, however, are made of quartz, glass not being transparent to the full UV spectrum. Amplifiers are sometimes included within the sensor to boost the signal, which is low since radiation levels are low (being confined to a narrow band). The units used for UV measurement are the same as for other solar radiation bands, that is watts per square metre, a typical maximum sensor coverage being around 50 W m^{-2}. Accuracy ranges from $\pm 2\%$ to $\pm 7\%$.

Pyrheliometers

Pyrheliometers are specialised instruments for the measurement of direct solar radiation at right angles to the incoming beam. There are several designs, but all are similar in that a thermal sensor is exposed at the bottom of a tube which is kept pointed at the sun. Using circular apertures, only radiation from the sun, and from a small annulus around it, is allowed to fall on the sensing surface, the angle of view to achieve this being 5.7°.

As with thermal solarimeters, it is the rise in temperature produced by the solar radiation falling on a black surface that is used to measure intensity, the black surface being at the bottom of the tube. It is mostly the means of measuring the rise of temperature that differentiates the various designs of sensor. One design is very similar to a normal solarimeter: the temperature of a black disc relative to the body of the tube is measured with a thermopile, the tube being kept at ambient air temperature (by virtue of being white or of polished chromium). It differs from a solarimeter simply in that the receiving disc sees only the sun's disc and is at normal incidence to it.

In another design, the temperatures of two blackened strips of platinum are measured. One strip is exposed to the sun; the other is in shade but electrically heated, the energy required to bring it to the same temperature as the exposed strip being known by means of measurement of the current through it and the voltage across it. This energy is the same as the energy being received from the sun.

For meteorological synoptic (forecasting) stations, readings are made man- ually at true solar noon and at times when the sun's angle from the zenith is such that the length of the path of the radiation through the atmosphere is 2, 3, 4 and 5 times the thickness of the atmosphere vertically above mean sea level.

At these times, the observer must also make judgements regarding the sky's whiteness or blueness, since a variety of aerosols is often present, attenuating and scattering the beam. There are rules for judging the state of haziness of the sky, but the process is partly subjective. Pyrheliometers can also be operated continuously, the sensor being driven by a motor so that it follows the passage of the sun across the sky. Measurements are recorded on a chart or logger, although information on the haziness of the sky will not be available from automatic sites. Most instruments can have a filter placed at the end of the tube in the path of the radiation, allowing specific bands to be measured. Normally the tube is capped with plain glass.

The accuracy of pyrheliometers is similar to that of pyranometers since they are similar in principle. However, as they are always operated at normal incidence to the sun's beam, there is no cosine or azimuth error.

Net pyrradiometers

The measurement of net radiation is particularly important because it gives information on the amount of energy available to heat the atmosphere and the ground, to evaporate water and to power photosynthesis and transpiration.

A *net pyrradiometer* measures short- and long-wave radiation from about 0.3 to 70 μm, the heating of a black surface again being the sensing method. One black surface faces horizontally down, another up; a thermopile measures the difference in their temperatures and thus the energy exchange (Fig. 2.9).

The protecting domes cannot be of glass as this is opaque to the longer-wave radiation. Instead polyethylene, which transmits the full solar and terrestrial spectrum, is used. Such domes are not as strong as glass and so are more vulnerable to damage (for example by birds) and they also age, owing to sunlight and weathering. Furthermore, they are less easy to keep clean and there is the problem that water gets into the cavity, both at the seal that holds the domes to the case and also because the material is slightly pervious to water vapour. To protect against this, an interchangeable silica-gel dessicating tube is built into the case; through this the cavity breathes. In an alternative design, thinner polyethylene domes are used and are kept inflated by purging them with a slow flow of dry nitrogen gas, but this is inconvenient owing to the need for gas cylinders and has been largely dropped in favour of smaller, more rigid domes. A heating ring can be fixed around the side of the domes, invisible to the sensing surfaces, to keep them dew- and frost-free. However, one of the difficulties with measuring net radiation is that the domes and the body of the instrument emit infrared radiation themselves, just as in the case of the glass domes of solarimeters, and this can introduce errors.

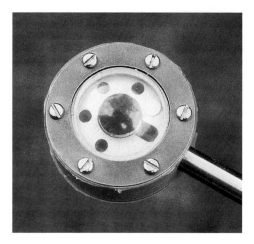

Figure 2.9. A net radiometer. Its upper black surface, protected by a dome of polyethylene (which is transparent to both solar and terrestrial radiation), senses the total incoming radiation. An identical, downward-looking, surface senses the total outgoing radiation, the sensor output being the difference between the two, the net energy exchange.

In an attempt to overcome this problem of radiation from the domes, there have been designs which dispense with them completely, the black surfaces being exposed directly to the radiation (Latimer 1962, Dept of Transport, Canada 1962). But this leaves the surfaces open to the weather and so, to offset the effect of varying cooling by wind, they are forced-ventilated by a motor-driven fan. However, not only does this require electrical power for the motor, it also requires corrections to be made. Also during and after rain the sensors are inoperative, since the upper surface becomes wet and cooled. For these reasons, this type of sensor is not much used today, although recently a net sensor without domes and without ventilation became available commercially; it comprises two slightly conical black surfaces. It is not intended for precise work since rain and wind affect its response, tests apparently indicating an error of 0.8% in the radiometer reading for each metre per second increase in wind speed. Even sensors with domes are to some, but a lesser, extent affected by water on the domes, since this affects the passage of radiation and itself radiates in the infrared. There is also some slight dependence on the wind speed over the domes.

Effective pyranometers

Also known as total radiometers, these instruments constitute half of a net radiometer and measure the combined total input of energy across the full radiation spectrum from (nominally) 0.3 to 100 μm. They can be either up-ward-facing, measuring the total solar input (direct and diffuse) together with the terrestrial radiation from the atmosphere and clouds, or downward-facing, measuring the reflected solar radiation together with the long-wave infrared emissions from the ground. The cold junction of the thermopile is the instrument case, kept at ambient air temperature by a shield. Sensors of this type are not widely used.

Pyrgeometers

Pyrgeometers are sensors that measure just long-wave radiation of terrestrial origin in the spectral band from 3 to 100 μm, and are simply pyranometers modified by changing the glass covers to domes that are opaque to the main part of the solar spectrum but transparent to the longer wavelengths arising from absorption and re-emission. Otherwise they are similar, in that a black surface is warmed through exposure to the radiation, its temperature relative to the case being measured by a thermopile.

By using two pyrgeometers, one facing down and the other up (in a way similar to that in which two pyranometers are used for the measurement of albedo), in combination with an albedo sensor, the same information is obtained as from a net radiometer, since the one pair measures net long-wave energy exchange and the other net short-wave. This combination provides a comprehensive measurement of the radiation environment – incoming and reflected solar radiation, albedo, incoming and outgoing terrestrial radiation and total net exchange of energy.

Exposure of radiation sensors

The Campbell–Stokes sunshine recorder requires a clear view of the sky, with nothing that will cast a shadow on the sensor wherever the sun may be at any time of the day or year, although below about 3° the sun has insufficient energy to burn a trace and so visibility right down to the horizon is not essential. Exposure on the top of a building may actually help in this case, although for most other sensors exposure on a roof is definitely not to be recommended, as will be shown. In the case of electrical radiation sensors, the exposure rules are the same, although in addition there should be no bright reflecting object nearby.

The same criteria must be met by net radiometers; furthermore, because they also measure the outgoing radiation the ground beneath them needs careful attention. The ground should be as representative of the area as possible,

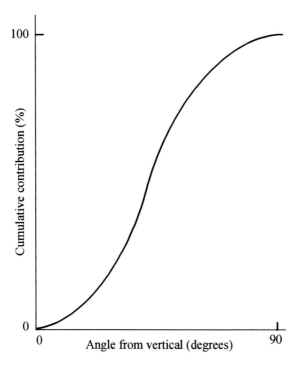

Figure 2.10. The proportion of irradiance received by the downward-looking sensing surface of a net radiometer (or of an albedo sensor) varies with the angle, the proportion of the contribution peaking at 45 degrees; the graph here shows the percentage cumulative contribution for increasing angles from the vertical.

whether it be vegetation, soil, rock, snow or water, since its reflective and infrared-radiative properties affect the measurements.

Net sensors are usually mounted at a height of from one to two metres. Figure 2.10 illustrates the cumulative contribution of radiation received from the ground for inceasing angles, from which it will be seen that 50% of the total input is received from a cone of semi-angle 45°. In the case of a sensor at 1 m height, this is equivalent to a circle of ground 1 m in radius (3.142 m²). Raised to 2 m, the area contributing 50% of the input increases by a factor of four. From the graph it can also be seen that 90% of the input originates within a cone of semi-angle about 65°, that is from a circle of 4.4 m radius (60 m²) if the sensor is deployed at 2 m height. Nothing is to be lost, therefore, by exposing the net sensor at two or more metres above the ground and, indeed, as raising the sensor increases the ground encompassed in its view, the measurement becomes more representative the higher the sensor is installed. The ground should, however, be representative up to the maximum distances included, and the mast which supports the sensor should be as unobtrusive as possible.

References

Amano, I., Koseki, K. & Tamiya, K. (1972) Photoelectric Shiko-type sunshine recorder. *Japan Meteorological Agency, Weather Serv. Bull.*, **39**, 77–9.

Burroughs, W. J. (1994) Weather cycles. *Weather*, **49**, 202–8.

Chappell, A., ed. (1976) Optoelectronics: theory and practice. Texas Instruments Ltd. ISBN 0 904 047 19 9.

Latimer, J.R. (1962) Laboratory and field studies of the properties of radiation instruments. Meteorological Branch, Dept of Transport, Canada, Circular 3672, TEC 414.

Meteorological Branch, Dept of Transport, Canada (1962) Manual of radiation instruments and observations. Circular 3812 INS 117, Manual 84.

Monteith, J. L. (1975) *Principles of Environmental Physics*. Edward Arnold, London.

3

Temperature

The variable

Only about 17% of solar radiation is absorbed directly by the atmosphere as it passes through it (Lockwood 1974). In fact the atmosphere is heated primarily as follows: solar radiation heats the ground and the ground's heat is transferred to the air, firstly by molecular diffusion across the *laminar boundary layer* (a layer only a millimetre or so thick, which clings to most surfaces); beyond this, in the *turbulent boundary layer*, transfer is by turbulence, which is much more effective at transferring heat than is diffusion. Heat is also, thereafter, transferred by convection, bubbles of warmer air rising into the cooler air above. This transfer of warmed air away from the surface is the *sensible heat flux*. While it is difficult to measure the rate of energy transfer (see eddy correlation, Chapter 6), the resultant changes in air temperature are important and more easily measured.

A proportion of the heat from the warmed ground is also transferred downwards as the *soil heat flux*, the rate of transfer being influenced by the amount of water in the soil and also by the pore and particle sizes and the presence of vegetative material. Many biological processes are influenced by soil temperature, from the activity of micro-organisms to the germination of seeds and plant growth, and observations of soil temperature are made at depths down to a metre or more.

The importance of the oceans in weather and climatological processes is now firmly established, and in this the sea-surface temperature is of particular significance. While the measurement of sea temperature is made with identical thermometers to those used for the air, how the observations are made is dealt with in Chapter 13 on the oceans.

Scales and reference points

The first measurement of temperature dates from the time of Galileo (1564–1642), when primitive thermometers, open to the air and using water and other liquids such as spirit of wine and linseed oil, were made. It was not until 1720 that Fahrenheit used mercury, although the upper end of the tube still remained open. Only later was the end sealed and the air removed, giving the modern form of thermometer.

Various scales have been adopted, initially only one point being fixed, the rest of the scale being equal fractions of the volume. Many different fixed points were used. Fahrenheit later introduced the idea of two fixed points, his original points being the lowest point observed in Danzig in 1709 and body temperature, which for reasons unknown he set at 96°. This scale gave the melting point of ice as 32° and the boiling point of water as 212°. These two points were later used as the fixed points, when it had been established that they were constant and easily reproduced. In 1730, Reaumur noted that a volume of 1000 units of mercury at ice point became 1080 at the boiling point of water and so marked his scale from 0 to 80.

In 1736 the Swedish astronomer Anders Celsius adopted the convention of 0° for the boiling point of water and of 100° for ice point, thereby creating the centigrade scale. This convention was later inverted, giving the present-day zero for ice point. This reversal is attributed to different people, one being the Swedish botanist Linnaeus, another the Frenchman Pierre Christin. The centigrade scale is now called the Celsius scale in honour of its inventor. Since 100 degrees Celsius $= 212 - 32 = 180$ degrees Fahrenheit, $1\,°F = 5/9\,°C$.

In 1848 the Scottish physicist William Thompson, later Baron Kelvin of Largs, proposed the *absolute* or *thermodynamic* scale of temperature, in which absolute zero corresponds to the temperature at which all molecular motion stops. On the Kelvin scale, one degree Kelvin (known as a kelvin, K) is equal to one degree Celsius and absolute zero (0 K) corresponds to -273.15 on the Celsius scale. The melting point of ice is thus $+273.15$ K and the boiling point of water is $+373.15$ K.

For precise calibration purposes, the triple point of water (see below) is used as one of the fixed points, the boiling point of water (equilibrium between the liquid and vapour phases of water) at a standard pressure of 1013.25 mb being another. The triple point of water, $0.01\,°C$ (or 273.16 K) is produced for calibration purposes in an arrangement of two concentric glass tubes. The outer is sealed and contains liquid water, ice and water vapour in equilibrium. The inner tube, into which the thermometer under test is placed, is open to the atmosphere and contains water in equilibrium with the surrounding outer tube at 0.01 °C.

The scale of temperature used today is the *International Practical Temperature Scale* (IPTS) agreed in 1968. It is based on a number of equilibrium states, including the triple point of water and the boiling point of water (at standard atmospheric pressure) and also the boiling point of oxygen (-182.962 °C or 90.188 K) at standard atmospheric pressure. There are also other secondary reference points, one of which is the familiar ice point of water (0 °C). Although called the IPTS, it is in practice the familiar Celsius scale.

Mechanical thermometers

Liquid-in-glass

Present temperature   Any property that changes with temperature can be used as the basis for a thermometer, the most common in use still being, as in Galileo's time, the thermal expansion of a liquid in a glass tube. As later introduced by Fahrenheit, the liquid is still usually mercury, but below its freezing point of -38.8 °C ethyl alcohol is used instead, which freezes at -115 °C. Although the coefficient of expansion of alcohol is much greater than that of mercury, it is not so effective as a thermometric liquid because its composition changes slowly with time (being organic). It also adheres to the glass, particularly if the temperature changes quickly, and the liquid column tends to break up. Alcohol also responds more slowly than mercury because its thermal conductivity is lower.

Liquid-in-glass thermometers make use of the fact that the liquid expands with temperature more than its glass container and so is forced up a fine-bore stem from a bulb containing the bulk of the mercury. For most meteorological and other environmental applications, the stem of the thermometer is graduated in steps of 0.5 °C from which the temperature can, with care and practice, be estimated by eye to 0.1 °C (Fig. 3.1). For periodic checking, thermometers are compared with a precise standard, graduated in steps of 0.1 °C.

Meteorological thermometers are manufactured to within certain tolerances. For mercury, from -40 to 0 °C the maximum error is typically between -0.3 °C and $+0.2$ °C while from 0 °C upwards it is between -0.2 and $+0.05$ °C. For spirit thermometers, below -40 °C the maximum error is ± 0.06 °C, from -40 to 0 °C and from $+25$ °C upwards it is ± 0.25 °C, while from 0 to $+25$ °C it is ± 0.1 °C. Readings may be in error anywhere within these limits. The figures are typical UK values and may vary slightly from country to country and from manufacturer to manufacturer.

The two main sources of error in mercury-in-glass thermometers are, firstly,

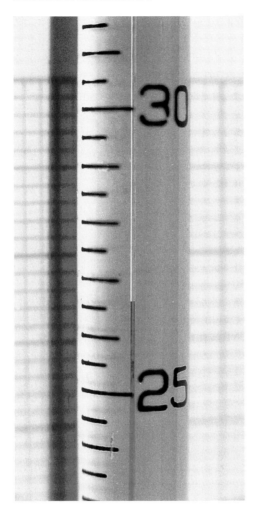

Figure 3.1. Although ordinary mercury-in-glass thermometers are graduated in 0.5 degree steps, tenths of a degree can be estimated; in the photograph the reading is 26.6 degrees.

changes in the glass of the bulb, which contracts slowly over the years introducing a rise in the zero (and so also in the readings) of about 0.01°C per year initially, although it slows up later. Secondly, if the liquid column is not read at eye level, a parallax error is introduced which can be up to ±0.2°C. Small breaks in the mercury column can also form, resulting in a higher reading.

A mercury-in-glass thermometer has a *time constant* (the time to reach 63% of the final value after a step change in the temperature) of about one minute for a bulb of 10 mm diameter in an airstream of $5 \, m \, s^{-1}$. In practice, temperatures tend to be cyclic and uneven and it is difficult to predict theoretically how a thermometer will behave under such conditions, but any temperature fluctuations faster than the time constant will not be registered. When taking a reading of the present temperature it is desirable that the thermometer does not respond too quickly, as it would then be unduly affected by short-term variations. The average over a period of about 5 to 10 minutes is of the right order, and for this a time constant of no less than about 1 minute is appropriate.

Maximum thermometers   Thermometers able to hold the maximum reading since last read are installed in most temperature screens worldwide. They are similar to the ordinary mercury-in-glass thermometer except that there is a constriction in the mercury column between the bulb and the lowest gradu-ation on the stem (Fig. 3.2). As the temperature falls, the mercury column breaks at the constriction and is not drawn back into the bulb, thereby holding its position. If the temperature rises again and to a higher value, the column is reformed and the reading rises to the new high. It is then held at this new point when the temperature again falls. To prevent gravity from moving the de-tached mercury column, the thermometer is operated on its side. It is reset by a sharp shake downwards, as with a clinical thermometer. Tolerances of manu-facture, errors and resolution are the same as for the ordinary type.

Minimum thermometers   Alongside a maximum thermometer there is usually one for recording the minimum temperature. This differs from the ordinary and maximum types in that it uses clear spirit as the liquid. Into the liquid is introduced a slide indicator, which is free to move along the bore but which stops when it gets to the liquid's meniscus. The thermometer is set by inverting it so that the indicator falls and rests on the meniscus; it is then laid on its side. If the temperature rises, the slide stays where it is, the liquid going round it (Fig. 3.3). If it falls, the meniscus pulls the slide back towards the bulb, the slide being unable to break through the meniscus owing to surface tension. The minimum temperature is where the top of the slider rests. This thermometer suffers from the errors mentioned earlier for ordinary spirit-in-glass thermometers.

Mean temperatures   Maximum and minimum readings can be used to derive an approximate daily mean temperature by taking $\frac{1}{2}$(max. + min.). At stations where spot readings are also taken several times a day, such as every three hours for weather forecasting, a better mean can be derived by using $\frac{1}{8}$(03 + 06 + 09 + $\cdots$ + 24), where the numbers inside the parentheses refer to the temperature at the time of the spot reading. There are many variations on this, such as $\frac{1}{12}$(02 + 04 + $\cdots$ + 24) or $\frac{1}{4}$(02 + 08 + 14 + 20), each giving a different accuracy. The variation in the way the mean is obtained presents difficulties when comparing mean temperatures from different stations or

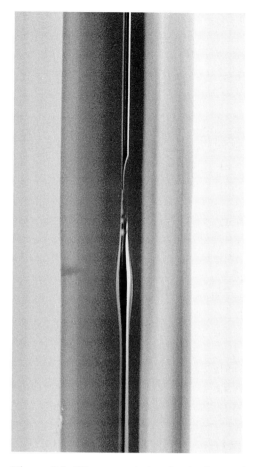

Figure 3.2. The constriction in the bore of a maximum thermometer breaks the mercury column if the temperature drops, preventing the column falling back below the highest reading. It must then be manually reset by shaking it.

when attempting to derive a global mean, as in climate studies (see Chapter 15). It is also not uncommon for the method used to calculate the mean to be changed during the lifetime of a station, further complicating the situation.

Combined maximum and minimum thermometers A Six's thermometer (Fig. 3.4) consists of a U-shaped tube with a bulb at each end, one being completely filled with spirit, the other only partly filled. The bottom part of the tube is filled with mercury. When the temperature rises, the liquid in the full bulb expands and forces the mercury towards the partly filled bulb, the reverse occurring when the temperature falls. Both ends of the mercury column indicate the present temperature. In the clear liquid above the mercury, on both arms of the tube, is a movable index, similar to that in a minimum

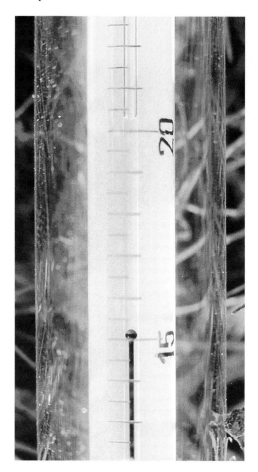

Figure 3.3. A minimum temperature of 15.2 degrees indicated by the sliding index of a minimum thermometer, the current temperature of 20.3 degrees being shown by the spirit meniscus. Although this thermometer is measuring minimum ground temperature, similar instruments are operated in Stevenson screens (Fig. 4.1).

thermometer. As the mercury moves backwards and forwards the indexes are pushed upwards on both sides. They have springs to stop them falling downwards when the mercury moves away, so that they stay at their highest points, which correspond to the maximum and minimum temperatures. The indexes are of steel, and are reset by drawing them down to the mercury with a magnet. This type of thermometer is of lower accuracy than the individual maximum and minimum thermometers because the graduations are not on the tube but on a backing plate to which the tube is fixed and also they are only marked in 1 °C steps. They are suitable for the amateur meteorologist and gardener, but not for accurate work.

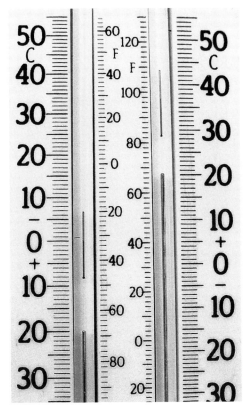

Figure 3.4. The two sides of a Six's thermometer, the lower ends of the index slides indicating a maximum of 28.5 degrees (right) and a minimum of 8.5 degrees, the present temperature, indicated by both ends of the mercury column, being 20 degrees.

Bimetallic thermographs

In principle these are like bimetallic actinographs for solar radiation measurement. The sensitivity of a bimetallic strip is greatest when it is straight, although for compactness in a thermograph it is usually coiled into a helix (for measuring solar radiation it is of necessity short and straight). Sensitivity is inversely proportional to the thickness of the strip and proportional to the square of its length and to the difference in the coefficients of expansion of the two metals. New strips suffer an aging process and so are subjected to repeated heating and cooling cycles before use. Sudden large changes, or temperature changes beyond their design limit, can permanently change the sensitivity and zero of bimetallic thermographs or damage them.

One end of the helix is fixed to the instrument's case, in such a way that the zero point can be adjusted by rotation of the helix. The temperature range is

Figure 3.5. The spiral bimetallic strip of this thermograph (outside the case on the right) indicates a fairly steady 10 degrees on the pen trace of the paper chart.

adjustable by altering the length of the helix. The other end of the helix is free to rotate, moving an arm holding a pen across a paper chart (Fig. 3.5). A spring- or battery-operated drum rotates the chart, of which there are several designs to suit different seasons and climates.

The time constant of this type of instrument is about 20 seconds, compared with the 50 to 60 seconds of a mercury-in-glass thermometer. Friction of the bearings is the main source of error and can be minimised by careful initial alignment of the mechanism and subsequent regular maintenance. The effect of the friction of the pen on the chart can be minimised if the bimetallic strip is made fairly rigid, thereby imparting its bending action more positively. A thermograph can, at best, be expected to have an accuracy of $\pm0.5\,^{\circ}$C but, to achieve this, care is necessary in making the initial settings and any necessary periodic adjustments. In practice it is possible that the readings may be out by considerably more than $0.5\,^{\circ}$C.

Electrical thermometers

Resistance thermometers

Pure metals   That conductors change in resistance as the temperature changes and can, therefore, be used to measure temperature has been known since the nineteenth century. Various metals and alloys can be used, the choice depending on such factors as the required stability over a long period, the uniformity of product and the degree of change in resistance with temperature and cost. All desirable features do not occur in one metal, but the most commonly used is platinum because of its stability and relative linearity; nickel is often a second choice when cheapness is important. Although all metals have a nearly straight-line temperature response they are not completely linear and for precise work a correction may be necessary.

Sensitivity to temperature change is defined by the material's *temperature coefficient*. For platinum this is 3.9×10^{-3} K^{-1} while for nickel it is 6.8×10^{-3} K^{-1}. Taking the example of a *platinum resistance thermometer* (PRT) with a resistance of 100 ohms at 0°C (the usual standard today and the most-used metal), the increase in resistance for a 10 degree rise is $100 \times 3.9 \times 10^{-3} \times 10$ (that is, 3.9 ohms) giving a total resistance at $+10$°C of 103.9 ohms, and so on, proportionally, for increasing and decreasing temperatures. Grade 'A' PRTs have a manufacturing tolerance of ± 0.15°C from -50 to 0°C increasing to ± 0.25°C at 100°C.

To measure the change in resistance, the thermometer can be made to be one arm of a Wheatstone bridge, or a constant current can be passed through the sensor, the voltage across it being measured. In both methods current flows through the sensor during the measurement, thereby heating it. Provided that the current is kept low, that the readings are made quickly and that there is an adequate flow of air past the sensor, the heating effect can be ignored. In the case where a data logger is used, the readings are taken in milliseconds and so heating is negligible.

Although resistance thermometers are now mostly used in automatic instruments, they have also been used at manual stations; the Wheatstone bridge is then balanced by the operator. This takes time and so self-heating is more of a concern, but the advantage is that the temperature can be read remotely without having to go to the screen and open it.

When the sensor is some distance from the measuring equipment, be it manual or automatic, it is necessary to compensate for the resistance of the wires between the sensor and the bridge, which would otherwise unbalance the bridge as well as introduce unwanted additional resistance changes due to temperature. Compensation can be achieved by including extra leads in the opposite side of the bridge to balance out those of the sensor's or, alternatively, the current can be fed along one pair of wires while the signal voltage across the sensor is picked off directly by another pair of wires that carry no current. A further complication of the low resistance of PRTs is that contact resistance within plugs and sockets can become a source of error.

It is possible to use software in a data logger to detect and store maximum and minimum temperatures as well as average or instantaneous values, the one PRT supplying all the readings.

Semiconductor temperature sensors   There is an alternative type of resistance thermometer to the PRT – the *thermistor*. This is a semiconductor heat-sensitive resistor that has a temperature coefficient an order of magnitude greater than those of pure metals. Sensors with both positive and negative slopes can be made and all types have a much higher value of resistance than PRTs. In addition they can be made very small, which is useful in certain applications requiring rapid response times or minimal radiative heating. The higher resistance of the sensors compared with a PRT (typically ranging from 3 to 100 kilohms at 20°C) eases the problem of lead resistance and of plug and socket contact resistance.

The non-linear response of thermistors is a slight disadvantage but it can be corrected by the use of two- or three-element composite thermistors within the probe. Or corrections can be made in the logger circuits or in the computer when processing the readings. Depending on how the linearisation is done and how the corrections are made, the accuracy of a thermistor varies from about ± 0.05 to ± 0.5°C. An aging process occurs and to reduce this thermistors are temperature-cycled by the manufacturer, keeping the long-term drift typically within the range of ± 0.015°C per year.

Cheaper thermistors may vary in resistance, one to another, and so are not directly interchangeable without recalibration. (Curve-matched sensors are, however, available at little extra cost, and these can be directly interchanged.) The same problem occurs with the cheaper PRTs. Figure 3.6 illustrates three thermistors alongside a miniature PRT; the pin gives the scale.

Thermocouples

In 1821, Seebeck discovered that where two different metals touch, a small voltage is generated. If two such junctions are formed and one is kept at a temperature different from the other, the consequent difference in potential between the junctions is proportional to their difference in temperature. This is known as a *thermocouple* and can be used to measure temperature difference. If a number of such junctions are connected in series, the voltage is added and the combination is known as a *thermopile*.

Because thermocouples measure difference in temperature rather than absolute temperature, they are well suited to applications such as radiation sensors (Chapter 2) and the Bowen-ratio method of measuring evaporation (Chapter

Figure 3.6. A pin illustrates the small size of three thermistors and a platinum resistance thermometer. The platinum is not in the form of wire wound on a bobbin, but is deposited on the plastic wafer as a fine track, its resistance trimmed by laser.

6). To measure absolute air temperature, however, it would be necessary to hold one set of junctions at a known temperature (for example $0°C$) and this is not very practical in the field. Thermocouples also produce rather low output voltages and so require higher amplification than the resistive types of temperature sensor.

Exposure of thermometers

Air temperature

Stevenson screens While solar radiation passes through the atmosphere without much absorption, a thermometer exposed to the same radiation will heat up considerably, the extent of the heating depending on the colour, texture and size of the thermometer, but the temperature rise can be up to $25°C$ above air temperature. Only in the case of very small thermistors, of PRTs made of fine wire or of thermocouples will radiative heating be negligible.

So, in measuring air temperature it is necessary to shield whatever type of thermometer is used from both solar and terrestrial radiation, and also from precipitation, while allowing air to pass freely into and through the screen from outside without any change in its temperature. The inside surfaces of the screen should also be as near as possible to the outside air temperature. The first person to realise all this and to design a suitable screen was the Scottish lighthouse engineer Thomas Stevenson (father of Robert Louis Stevenson), who, in 1866, designed the prototype of the now familiar wooden temperature screen (Figs. 3.7 and 4.1). These wooden structures have double-louvred sides to provide insulation from the outer body, with an opening front to give access to the thermometers. The roof is also double, with an air space, while the base is made of overlapping boards to allow free movement of air.

Standard screens have internal dimensions of about 1 m width, 0.5 m height and 0.3 m depth, which is large enough to contain a variety of thermometers as well as a bimetallic thermograph and a hygrograph. There are larger screens, and a half-width model is used to contain just the liquid-in-glass thermometers. On board ships where there is little room, a smaller screen with single-louvred walls is used, just big enough to hold two thermometers (wet and dry for humidity measurement). However, worldwide, there is a variety of designs, some bigger than those described above.

Little work has been done on the time constants of wooden screens, but the work of Langlo (1949), Bryant (1968) and Painter (1977) points to values of from 4 to 17 minutes for wind speeds of 10 to 0.5 m s^{-1}, respectively. As regards the effectiveness of the screen at shielding the thermometers from radiation and of ensuring that the air inside the screen is as close as possible in temperature to that outside, Painter found that, in the UK, if the conditions were unfavourable (bright sun and low wind speed) the temperature could be in error by +2.5°C, while on a cloudless calm night the screen could be colder by up to −0.5°C. It is important also to recognise that if the screen is not kept clean it will be less effective, and this will cause additional error. Repainting should be done every two years and the screen should be washed on alternate years. Keil (1996) showed that temperatures could be lower by 1°C after the repainting of an old screen. These factors are likely to be even more of a problem in climates which have many hours of sunshine and high solar radiation intensities.

The opening of the front of a screen to take readings exposes the thermometers to radiation from outside. Readings must, therefore, be taken quickly to avoid error, but not so hurriedly that mistakes are made. Since automatic weather station (AWS) temperature screens are not opened to take readings, this source of error is avoided (see below).

The best site for a screen is at the most exposed position available. The top of

Figure 3.7. The open door of this Stevenson screen exposes its louvred construction and the typical arrangement inside of maximum and minimum (centre) and wet and dry thermometers (see Fig. 4.1 for close-ups) along with a bimetallic thermograph (on the right, see Fig 3.5) and hair hygrograph (on the left).

a building is not, however, appropriate, since temperatures can change with height quite considerably, especially at night, and also because the building itself will have a great influence on the temperature. Nor, for example, is a hollow in the landscape or a steep slope suitable, as it may well have an untypical microclimate. It is traditional to operate screens over short-cut grass, but obviously if the surrounding country which is to be monitored is desert or rock or snow-covered or an expanse of water or dense forest, the screen should be deployed appropriately. Because quite considerable temperature gradients can occur near to the ground, it is important that screens are mounted at the standard height of 1.25 metres. Where this is not possible the height must be known so that corrections can be made. Over a snowpack that changes depth with season, it is usual to move the screen up and down to keep it the same distance above the snow.

Aspirated screens   The Stevenson screen is ventilated naturally, relying on the wind to move air through it. While this is generally good enough, it does introduce some error. To overcome the problem of varying air speed, for precise work, aspirated screens are used in which the air is circulated past the sensors with a fan at a known speed. The commonest type of power-ventilated screen was designed for intermittent, manual operation, that is for taking spot measurements. It was also intended primarily for the measurement of relative humidity using the wet-and-dry-bulb method, where a good passage of air is even more essential, and so it is described in the next chapter, on humidity.

Aspirated screens provide one of the most accurate ways of measuring air temperature and, if higher precision is required, aspiration is sometimes used at automatic stations (if there is sufficient electrical power to drive a fan) and a number of designs are made for this purpose. For more precise results, Stevenson screens are also occasionally artificially ventilated, a fan on top drawing in the air. There are also naturally aspirated screens that turn into the wind to make the best use of any natural ventilation.

The whirling hygrometer provides ventilation manually, by whirling the thermometers around on a frame, with a handle. It has no screen and has to be kept out of direct sunlight and read quickly. It too is dealt with in the next chapter since it is intended essentially for measuring humidity.

Small screens for automatic weather stations   In the early days of AWSs (the 1960s) electrical thermometers were sometimes deployed in ordinary wooden Stevenson screens. While this had the advantage of some compatibility with manual systems, it was no longer actually necessary now that the sensors were smaller, and so the first miniature temperature screens were developed. A typical design is shown in Fig. 3.8. Mostly they are naturally ventilated, although aspirated designs (Fig. 3.9) are made for situations where power permits and where natural ventilation is always very low, for example in a dense forest.

The problem of changing from one type of instrument to a new one, as explained earlier, is that it introduces a step change in the observations, and this occurs if a change is made from Stevenson screens with liquid-in-glass thermometers to AWS screens with electrical resistance thermometers. Huband (1990) investigated the performance of three similar screens of the type shown in Fig. 3.8, comparing them with each other and with a Stevenson screen. What he found was that measurements in the AWS screens agreed with those made in the Stevenson screen to within about $\pm 0.5\,^{\circ}$C and for much of the time to within $\pm 0.2\,^{\circ}$C, although the differences could at times be as much as $\pm 1\,^{\circ}$C, the latter probably being due to the different time constants of the screens since these larger differences occurred mostly at times of rapid temperature change. Maximum temperatures occurred within 10 minutes of each other but there was more delay in the Stevenson's minimums, probably owing to this screen's greater thermal mass. Earlier, McTaggart-Cowan & McKay (1976) had tested

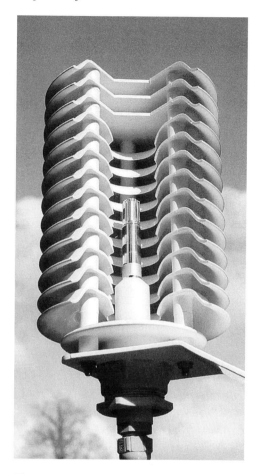

Figure 3.8. Automatic weather station (AWS) temperature screens are much smaller than the large wooden Stevenson screen, although they perform the same function of shielding the sensors from radiation. This shield is cut away to show a cross-section of the louvres. Inside is a small tube containing a miniature resistance thermometer and a humidity sensor (Fig. 4.5).

19 screen designs, including aspirated models, and while they found that the aspirated screens were within the same ± 0.5 and $\pm 0.2\,°C$ range of agreement, the non-aspirated were found to be less accurate, particularly with regard to maximum and minimum temperatures, each one being different.

Ground temperature

The same types of thermometers used for the measurement of air temperature are also used with little modification to measure ground temperature. It is largely how they are exposed that differs.

Figure 3.9. An AWS screen may be aspirated by an electric fan to obtain more precise measurements. This is more important where wet and dry bulbs are being measured to derive humidity. In the screen shown here, the small cylinder on the left contains the fan, the larger cylinder on the right containing the wet-bulb water supply.

Surface minimum temperatures   The minimum temperature reached overnight of a surface exposed to the sky is measured with the same type of minimum thermometer that is used for measuring minimum air temperatures in a Stevenson screen (Fig. 3.3). In this application, the thermometer is exposed without any screen, almost horizontally, with as clear and unobstructed a view of the sky as possible.

The minimum temperature of three types of surface is measured: short-cut grass, bare soil and concrete. Their radiative properties being different, they experience different minimums and thus give an indication of what radiative losses are occurring. As with most types of long-term observation, precise working practices have become established, both within national weather services (NWSs) such as the UK Meteorological Office and through the recommendations of the WMO. Measurement of ground-temperature minimums requires that the thermometer is almost horizontal, inclined at around 2°, the bulb just touching the tips of the short-cut grass, bare soil or concrete. The latter is usually a light-grey-coloured slab, 50 mm thick with sides of 1 m by 600 mm, its surface being set flush with the ground. If there is snow, the thermometers in the cases of grass or bare soil are laid on the snow surface, but not quite touching, while in the case of concrete, the slab is swept clear of snow when the thermometer is set. The usual practice is to put the thermometers out just before sunset, but at many sites they are left out permanently.

Because minimum thermometers use ethyl alcohol as the liquid, and suffer from the problem that the alcohol evaporates and recondenses further up the tube, minimum thermometers exposed on the ground often have their top ends

Figure 3.10. Soil temperature thermometers for depths of 5, 10 and 20 cm are normal mercury-in-glass instruments but with a right-angle bend in them.

covered in a black material that warms in the sun and so helps to prevent any evaporated liquid from condensing in the top of the tube, returning it instead to the main column of liquid.

The same measurements can be made using electrical thermometers, although at remote, unattended stations the sensors get covered if it snows. Logger software detects and stores the minimum readings.

Subsurface temperatures   In addition to the surface minimum temperatures, ground temperature is also measured at depths of 5, 10, 20, 30, 50 and 100 cm, although not necessarily at all these depths at all sites. There are three designs of mercury-in-glass thermometer for making ground temperature measurements.

Temperatures at 5, 10 and 20 cm depths are measured using a thermometer with a right-angled bend in it (Fig. 3.10) and with graduations of 0.5°C, allowing estimates to 0.1°C in the same way as for air temperature. Quality and manufacturing tolerance are also the same as for air thermometers. They have to be installed carefully so as to make good contact with the soil, and with minimum disturbance to it, so that readings are representative. Once installed they remain in place, undisturbed.

For depths of 30 cm and deeper, an ordinary thermometer is suspended in a steel tube of about 32 mm bore, set vertically in the ground. The tube is a permanent fixture, but the thermometer is withdrawn on a chain for reading, being installed in a glass case with rubber 'O' rings on the outside to protect it and also to minimise air convection within the tube. The thermometer bulb is embedded in wax (Fig. 3.11) to increase its time constant so that the reading does not change while it is out of the tube being read. This is acceptable because ground temperatures change very slowly.

Figure 3.11. For measuring soil temperature below 20 cm, a normal thermometer housed in a glass outer tube is lowered on a chain down a permanently installed metal tube. The cap that supports the thermometer chain
seals the tube at the top, while two 'O' rings on the glass tube, just contacting
the inside of the metal tube, prevent convection. The thermometer is kept at a
stable temperature, while temporarily withdrawn for reading, by being permanently sealed in a mass of wax (at the bottom of the photograph).

Figure 3.12. An alternative type of ground thermometer, for use at all depths, is similar to that for use down to 20 cm, except that the bend is not 90 but 30 degrees. Such a thermometer is left permanently in place.

The second, but less common, method uses thermometers similar to those for depths up to 20 cm, but with a 30° bend instead of 90°. These are used for all depths, right up to one metre (Fig. 3.12) and are left in place once installed. They are very vulnerable to breakage during installation and are difficult to handle. Better contact with the soil and the absence of a steel tube make for a less disturbed environment, although tests with the tube method indicate that in fact errors due to conductivity along the tube, or convection within it, are negligible.

It is usual to measure ground temperatures up to 20 cm depth under bare soil and temperatures at greater depths under short grass, whatever the type of thermometer used. But it does depend on circumstances, and grass would be inappropriate in, say, a desert.

Electrical resistance thermometers can be installed in the ground at any depth and connected to a data logger. They offer by far the simplest method of measuring soil temperatures and can be buried and left undisturbed, with only a thin wire emerging at the surface, thereby minimising any heat conductance from the surface and with the added advantages that automatic logging confers.

Soil heat flux   It is not just the temperature of the ground that is relevant; the flux of heat through the ground is also of importance since it is one indicator of the relative division of net radiation into sensible, latent and soil heat fluxes. Knowing the value of one of these fluxes helps to estimate the others. For measuring the flux of heat through the soil, a *heat-flux plate* is used. This is normally a disc about 10 cm diameter and a few millimetres thick, of a material that has a similar conductivity to soil. It is buried in the ground and the difference in temperature between the upper and lower surfaces is sensed by a thermopile; some sensors use thermistors.

If the sensor differs in conductivity from the soil, it will disturb the heat flux around it. It is also difficult to instal the plate without undue disturbance of the soil. To minimise disturbance, the sensor is sometimes installed by digging a pit and inserting the plate sideways. Even then it is difficult to ensure good thermal contact with the soil and it is also likely that the plate will move slightly with time. And even with careful back-filling, the pit will probably modify conditions and affect the flux (Gilman 1977).

References

Bryant, D. (1968) An investigation into the response of thermometer screens – the effect of wind speed on the lag time. *Met. Mag.* **97**, 183–6, 256.

Gilman, K. (1977) Movement of heat in soils. Institute of Hydrology Report No. 44.

Huband, N. D. S. (1990) Temperature and humidity measurements on automatic weather stations. A comparison of radiation shields. Internal report, Campbell Scientific Ltd.

Keil, M. (1996) Temperature measurements in a Stevenson screen. University of Reading, Dept of Meteorology report.

Langlo, K. (1949) The effects of the solar eclipse of July 1945 on the air temperature and an examination of the lag of the thermometer exposed in a screen. *Met. Ann. Oslo*, **3**, No. 3, 59–74.

Lockwood, G. L. (1974) *World Climatology, An Environmental Approach.* Edward Arnold, London. ISBN 0 7131 5701 1.

McTaggart-Cowan, J. D. & McKay, D. J. (1976) Radiation shields – an intercomparison. Canadian Atmospheric Environment Service unpublished report.

Painter, H. E. (1977) An analysis of the differences between dry-bulb temperatures obtained from an aspirated psychrometer and those from a naturally ventilated large thermometer screen at Kew Observatory. Meteorological Office, UK, unpublished report. Copy available in National Meteorological Library, Bracknell, UK.

4

Humidity

The variable

Just as air, warmed by contact with the ground, is transferred into the atmosphere by processes of diffusion, turbulence and convection, so too is the water vapour produced by evaporation. The ratio in which the net radiative energy is divided between heating the atmosphere, heating the ground and evaporating water is dependent on many factors, such as the amount of water actually available, the nature of the ground and the type of vegetation. Knowing the rate of evaporation of water is useful information in hydrology, meteorology and agriculture, but it is difficult to measure. However, the amount of water vapour in the air, i.e. the air's humidity, is easier to measure and this chapter looks at how it is done; Chapter 6 addresses the more difficult problem of how evaporation rates are measured.

Units and terminology

Hygrometry is the measurement of the water content of solids, liquids and gases; in environmental applications this usually means the water content of the atmosphere. Water vapour exerts a pressure, the *vapour pressure* (VP) of water, which can be measured in any of the usual units of pressure, such as millibars. Above a water surface in an air-filled container, the VP rises to a maximum level, the *saturation vapour pressure* (SVP), beyond which it cannot rise any further at that temperature (the rate at which water molecules are leaving the surface of the water being then the same as the rate at which they return due to molecular bombardment). The higher the temperature the higher the SVP: at 0°C it is 6.108 mb. There are equivalent SVPs over ice below 0°C. All these SVPs are given in hygrometric tables.

Cooling a sample of air (while keeping its pressure constant) increases its VP, and a temperature will be reached when the SVP is arrived at; further cooling then causes dew to form. This is known as the *dew-point* temperature. There is a similar temperature, below $0\,°C$, known as the *frost-point* temperature (or hoar-frost point).

The *relative humidity* (RH) is the ratio (expressed as a percentage) of VP to SVP. Thus 100% indicates saturation. This applies just to the temperature of the sample at that time so, without also knowing the air temperature, the RH is not complete information.

More academic are units such as the *mixing ratio* (the ratio of the mass of water vapour to the dry mass of the air it is associated with); the *specific humidity* or *water content* (the ratio of the mass of water to the mass of moist air); and the *vapour concentration* or *absolute humidity* (the ratio of the mass of water vapour to the volume of the moist air it is associated with), which is in effect the density of the water vapour, hence its additional name of *vapour density*.

Measurement techniques

The first recorded attempts to measure humidity date from around 1500 when Leonardo da Vinci described the effect of humidity on the weight of a ball of wool, while in 1665 Robert Hooke noted that the length of a gutstring changed with the humidity. All methods of humidity measurement, old and new, can be divided into four groups.

The addition or removal of water   Water vapour is added to or removed from the air, the resultant changes being measured. This category includes the gravimetric, volumetric and psychrometric methods. In the first two, water vapour is removed from the sample by a desiccant and the gain in weight of the desiccant, or the change in volume of the sample, is measured. These methods are laboratory techniques and are not described further. However, the psychrometric method, the wet-and-dry method, is one of the most practical and widely used methods of measuring humidity, and so is covered fully below.

The adsorption and absorption of water vapour   Water vapour is adsorbed (taken onto) or absorbed (taken into) many materials, to an extent dependent on the level of RH, causing them to change in some physical, chemical or electrical way, the processes being reversible and repeatable. By measuring the change, an indication of RH can be obtained. Several instruments use these properties, in particular the hair hygrometer, the thin-film capacitive sensor and the ion-exchange sensor, all of which are in wide use and so are fully described below.

The condensation of water vapour   One of the most precise ways of measuring humidity is by detecting the temperature at which dew forms. This class of instrument includes the dew-point and dew-cell hygrometers, but neither are widely used for everyday applications, although the dew-point method provides a good laboratory standard and finds application in the Bowen-ratio method of measuring evaporation.

The radiation absorption method   Measurement of the absorption of infrared radiation passed through air containing water vapour allows rapid changes of humidity to be sensed, but it is little used except for specialised applications such as the eddy correlation measurement of evaporation.

The psychrometric method

When evaporation takes place from a wet surface into an airstream, the surface cools until it reaches a state of equilibrium, the amount of cooling depending on the relative humidity, the air temperature and the airspeed. By measuring the temperatures of the air and of the wet surface, the VP can be determined. This is the wet-and-dry method and is one of the most commonly used. It is simple, cheap, works from very low humidities to 100% RH, is precise and consumes no power.

Basic principles

A *psychrometer* consists of two thermometers (mercury or electrical), one measuring air temperature and the other, kept wet by a wick dipped into distilled water, measuring the amount of cooling produced by the evaporation of water from it, which is dependent on the relative humidity.

The psychrometric formula $e = e_s - Ap\,(T - T_w)$, relates the vapour pressure e, the air temperature T, the wet-bulb temperature T_w, the atmospheric pressure p, the saturation vapour pressure e_s (at temperature T_w) and the psychrometric coefficient A.

This coefficient has been derived both theoretically (Wylie 1968) and by experiment, but it is the one factor that cannot be entirely precisely quantified. However, over the decades its value has been carefully estimated for ordinary mercury-in-glass thermometers, and with air speeds of $3\,\mathrm{m\ s}^{-1}$ or more the coefficient is in the order of 0.667×10^{-3} K^{-1}. For smaller bulbs or greater ventilation rates, the coefficient decreases.

Below freezing point, the wet bulb becomes an ice bulb, although evaporation continues to occur (as regelation), and so the ice bulb is colder than the dry bulb and humidity can still be calculated – using different tables. However, the ice bulb will eventually dry out and will not then give meaningful readings. At an attended site, the bulb can be manually painted with water to make an ice bulb, and for this there are set procedures. This is not possible at automatic unattended stations.

Screens and accuracy

The accuracy of the wet-and-dry method is mostly dependent on the rate of flow of air past the wet bulb and this varies depending on the screen used.

Stevenson screens   The wet-and-dry method is widely used throughout the world, in Stevenson screens (Fig. 4.1). Indeed it is probably the most-used method for measuring atmospheric humidity. The ventilation is natural in this type of screen, introducing some uncertainty over the psychrometric constant. Folland (1975) investigated the problem and suggested that in a large screen the mean ventilation rate is around 0.75 m s^{-1}. Others (Bultot & Dupriez 1971) showed that only rarely did the rate reach 1 m s^{-1}. For natural ventilation of this order, the UK Meteorological Office uses a psychrometric coefficient of 0.799×10^{-3} K^{-1} above 0°C and 0.720×10^{-3} K^{-1} below freezing. Other inaccuracies arise if the wrong muslin is used to cover the wet bulb, if the muslin is dirty or contaminated, or if distilled water is not used. How well the wick fits the thermometer is also of importance and there are many rules regarding this, such as are given in *The Handbook of Meteorological Instruments* and *The Observer's Handbook* (Met. Office 1982a, b). If the dry bulb becomes wet, through fog or condensation, it will act as a wet bulb, with consequent errors.

Because of these many factors it is not possible to give a firm figure for the accuracy of humidity measurements in a Stevenson screen, but most manufacturers claim $\pm 1\%$ to $\pm 2\%$ RH, with the warning that this is dependent on air temperature and wet-bulb depression. There must, therefore, be some doubt as to the accuracy of humidity measurements from Stevenson screens, and $\pm 1\%$ to $\pm 2\%$ must be viewed as the best obtainable, under optimal conditions.

Aspirated screens   For more precise measurements, the air is circulated past the thermometers by a fan in an *aspirated psychrometer* (also known as an aspirated hygrometer or Assman psychrometer). A typical psychrometer con-

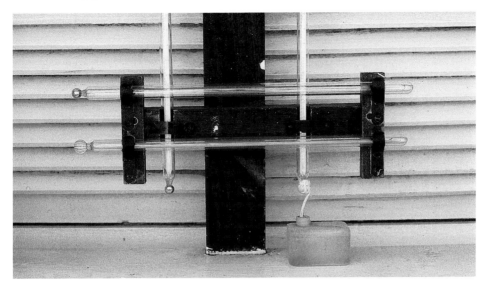

Figure 4.1. Inside most Stevenson screens are four basic thermometers – wet and dry bulbs (the vertical thermometers, right and left) and maximum and minimum thermometers (set horizontally, upper and lower). The wet-bulb wick is kept moist by the distilled water in the small plastic bottle. The difference in temperature between the wet and dry thermometers gives a measure of the relative humidity.

sists of a spring- or electric-motor-driven fan that draws air up a central tube, the tube being divided into two paths, one housing the dry thermometer and the other the wet thermometer. The wet bulb has a length of wick tied to it, but it does not dip into a water container as in a Stevenson screen, instead being wetted manually just prior to use. The fan is operated until stable readings are reached – after a few minutes. A built-in slide rule is then used to derive the relative humidity, dew point or vapour pressure from the wet-and-dry readings. The aspirated psychrometer is an accurate instrument and is often used as a standard against which to compare others.

A more economic way of obtaining aspirated wet-and-dry readings without either a screen or a fan is to mount the wet and dry thermometers on a simple frame. This arrangement is known as a *whirling* or *sling psychrometer* (Fig. 4.2), the whole construction being whirled around on a handle. Whirling at anything more than three revolutions a second produces the necessary wind speed and readings are taken at intervals until stable. The readings must be taken quickly since when movement stops the temperatures change quickly. It is also necessary to prevent the sun from shining on the instrument and if it is raining it must be keep under shelter. Used intelligently, good results can be achieved cheaply.

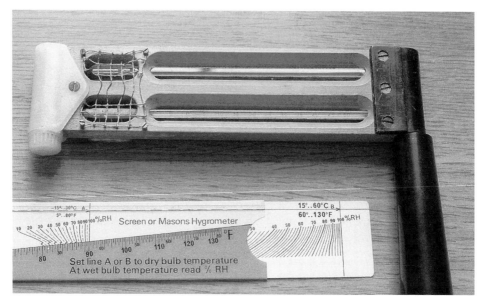

Figure 4.2. The whirling hygrometer is a simple and economic way of taking spot readings of air temperature and humidity; the wet and dry thermometers, mounted in a frame on a handle, are whirled by hand to aspirate them. The lower bulb in the figure is kept moist by the small water container (left), just as in the Stevenson screen. The slide rule enables humidity to be calculated without recourse to tables.

Because the ventilation is under control in an aspirated or whirling psychrometer, the psychrometric coefficient can be more precisely quantified, being taken as 0.666×10^{-3} K^{-1} above 0 °C and 0.594×10^{-3} K^{-1} below 0 °C, for any ventilation rate above 3.6 m s^{-1}. In such conditions, the humidity can probably be measured to within $\pm 1\%$ or $\pm 2\%$ RH. A few Stevenson screens have been equipped from time to time with fans to aspirate them. When this is done, a similar accuracy will be achieved.

AWS screens   Wet and dry electrical resistance thermometers can be used in automatic weather station (AWS) screens (Fig. 4.3). While this is a very effective method of measuring RH automatically, the AWS must be visited often enough to replenish the water, or an automatic method used for this; furthermore, if operated below freezing for any length of time, the data will be in error.

Although AWS screens are generally non-ventilated, being smaller than Stevenson screens their natural ventilation is probably better, and thus humidity measurements are likely to be somewhat more accurate than in a large wooden screen. However, while there are carefully prescribed rules specifying exactly how the wet-bulb wick should be fitted to a mercury-in-glass thermom-

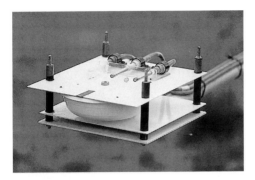

Figure 4.3. The wet-and-dry method is equally applicable to electrical thermometers, as in this early design of an AWS screen (top cover removed). The wet bulb (left) is kept moist by a wick that dips into the water container beneath. The dry thermometer (centre) is wired in a bridge circuit with the wet bulb to give a direct measure of the difference in temperature. The third thermometer (right) measures air temperature with a separate bridge circuit.

eter, there is a much greater variety of resistance thermometers in use, and no rules about fitting the wick, resulting in some uncertainty as to the exact value of the psychrometric coefficient. AWS screens are also much less well investigated than Stevenson screens and so the accuracy of humidity measurements made in them by the wet-and-dry method is not as certain. When used under poorly ventilated conditions, such as in a dense forest, or for more precise measurements, AWS screens may be ventilated by fans (Fig. 3.9). These will probably give a similar performance to other aspirated types, that is $\pm 1\%–2\%$.

The hair hygrograph

Many materials change their physical dimensions as the amount of water vapour in the air changes and such changes can be used to measure humidity. Human hair is one such material, and is widely used. (Others are cellophane, wool and gold-beater's skin, the latter being used in early radiosondes.) The hair hygrograph is almost as widely used as the wet-and-dry method, is well developed, well proven, reliable, and has a long and respectable history, despite its primitive concept and mechanisms. It is not, however, very precise.

Basic principles

Human hair, which is one of the most sensitive of the hygroscopic materials, has been used since the eighteenth century, when de Saussure made the first known hair hygrograph. As the RH changes from 0% to 100%, human hair increases in length by about 2.5%, in an approximately logarithmic way, most of the increase being at the lower humidity end. It is not much affected by any other vapours or gases, nor is its length very temperature dependent. The change in length is a function of the RH and not of the actual amount of water vapour in the air. To use this property, a band of hairs is kept under light tension, the change in length with changes in RH being magnified and linearised by levers and converted into a pen movement, which is recorded on a paper chart (Fig. 4.4).

Accuracy

Spilhaus (1935) found that there is a gradual drift in the length of hair held under light tension, affecting the zero setting. Although most of this can be corrected by periodically saturating the hair with distilled water, there is still a need to adjust the zero on a regular basis. Changes in the hair also affect the reading at the top end of the RH scale; this is checked regularly by covering the instrument with a saturated cloth, which produces a 100% RH atmosphere, and then making any necessary adjustment. The lower end of the scale is less simple to check since 0% RH is not so easily produced (although placing the instrument in a sealed container with sufficient desiccant such as silica gel will probably come close to achieving 0% if it is left there long enough). However, instead, the instrument can be put in a room having a steady temperature; when its reading is stable the RH is determined using an aspirated psychrometer.

The time constant of ordinary hair, above freezing point, is in the order of a minute, but at $-10\,°C$ it has risen to three minutes, at $-20\,°C$ to over six minutes and by $-30\,°C$ to about thirteen minutes or longer. If the hair is rolled flat under pressure, the time constant falls to no more than 30 seconds at $-30\,°C$ and to only 10 seconds at $30\,°C$, but the price paid is an increased brittleness of the hair and a consequent loss in its strength, causing it to break under tension more easily. The time constant is also dependent on the tension exerted on the the hair, on the previous treatment or history of the hair and on the ventilation rate, and it is different for increasing and decreasing humidities (hysteresis).

Manufacturers' claims vary as regards accuracy but a typical specification is $\pm5\%$ for 20%–80% RH. No claims are made for the top and bottom 20%. The hair hygrograph, however, is not intended as a precise instrument, its usefulness being its ability to make a chart recording of the changes in RH, but it does not do this to high accuracy.

Figure 4.4. The change in length with relative humidity of a bundle of strands of human hair, kept under light tension in a hair hygrograph, is transferred to a pen which records a trace on the rotating paper chart. The hair is here visible stretching from left to right, held in tension at its centre point.

Thin-film capacitive sensors

There are several types of *electrical humidity sensor* in which a thin film of material changes its capacitance or resistance in the presence of water vapour by an amount related to the RH. These are the most common type of electrical humidity sensor used today; most AWSs and hand-held instruments use them. Early designs were based on aluminium oxide (Jason 1965) or on thin polymer films (Nelson & Amdur 1965).

Basic principles

The most commonly used type of electrical humidity sensor is made by depositing a thin metal film in two sections on a glass substrate about 0.2 mm thick and about 5 mm square. Onto these two 'plates' is deposited a one-micron-thick amorphous organic polymer layer to act as the capacitor's dielectric. An upper single plate of water-permeable, vacuum-evaporated metal (palladium or gold), about 10^{-2} microns thick, is then formed on top of this (Fig. 4.5). The thickness of this upper plate is a compromise between the requirements that water vapour should pass through it quickly and that it should be thick enough not to have too high an electrical resistance. (The capacitor is operated in a tuned circuit and a low value of resistance in series with a capacitor leads to a high Q-value for the circuit – a high narrow peak output at the resonant frequency.) Because of the way in which it is constructed, with the lower plate in two sections and just one upper plate, the sensor

Figure 4.5. The 5 mm plates (dark square) of this thin-film humidity sensor change in electrical capacitance with changing relative humidity. In use it is protected by a permeable membrane in a metal tube, the whole being deployed in a screen (Fig. 3.8).

consists, in effect, of two capacitors in series. It is designed in this way so that electrical contact with the very thin upper plate is avoided (Stormbom 1995).

Water vapour molecules enter the polymer through the upper plate, forming bonds with the polymer molecules and thus increasing the electrical capacitance of the system; the capacitance is almost linearly related to RH (and not to the absolute amount of water vapour). The water molecules form two different types of bond with the polymer, one being very similar to those formed in liquid water, giving the water a dielectric constant of around 80 and thus a conveniently large capacitance change to measure. In the second type of bond with the polymer, the water molecules form dipoles with longer relaxation times. However, these two types of bond can be separated by the use of different frequencies to measure the sensor's capacitance. At radio frequencies, the capacitance changes induced by the liquid-water-like bonds predominate while the non-linear response of the second type only shows up at lower frequencies. Thus by using frequencies above about 1.5 MHz a linear response is obtained. (This is similar to the techniques used to measure the water content of soil; see Chapter 9.) The capacitance measured in this way is also relatively free from temperature dependence and no correction is necessary.

The water vapour absorption process occurs in two stages, the first taking less than a second since there are a fixed number of bonding sites in the polymer. But over a few minutes the number of sites increases owing to swelling in the polymer material and these become occupied more slowly, over minutes, the exact time and amplitude of the second stage of absorption depending on the particular polymer used. According to Salasmaa & Kostamo (1974) this two-stage absorption accounts for the hysteresis that some sensors of this type exhibit (there is no hysteresis in the short-term change). These time constants are dependent on temperature; below freezing point, the two effects having merged, there is no longer an initial rapid response.

Accuracy

Manufacturers claim that from 0% RH (in fact perhaps 10%–20% RH) to about 80% RH the thin-film capacitive sensor is accurate to about ±2% RH, and this is probably the case. From 80% to 100% it is specified as being ±3%. However, accuracy over the top 10%, especially at 100% and especially in cloud, must be open to some question. The long-term stability is also questionable, particularly if 100% is encountered for long periods. Performance at this upper end is expressed by manufacturer's statements such as 'Stability is better than 0.5% RH per year in normal air conditions (0%–70% RH)'. While this may be normal for offices and laboratories it is not the case out of doors, where 70% is regularly exceeded. If the sensor is operated in cloud or fog it can experience an RH of 100%+ (supersaturated), while in rain the RH approaches 96% – as it also does daily after sunset in many climates. Another manufacturer expresses it by saying that accuracy is ±3% above 80% if the sensor is kept continuously above 80% RH. which is impracticable out of doors. For use in cloud and in radiosondes, some of the latest sensors have built-in microheaters on the glass substrate to drive off dew or hoar frost; it is thereby acknowledged that at least in cloud some preventative measures are useful.

Ion-exchange sensors

An alternative type of electrical sensor is based on a conductive ion-exchange surface of chemically treated styrene copolymer (hence its alternative name of polyelectrolyte sensor). The water vapour molecules are adsorbed onto its surface, not absorbed into its thickness, resulting in a faster response time. However, the response time is shorter for increasing humidity, water vapour being released more slowly than it is taken up, with a time constant of about 30 seconds to one minute in the drying phase (depending on the rate of air flow). There is also a hysteresis effect: the difference is $\pm 3\%$ between the upward and downward excursions of RH. As with the capacitive sensors, there is a tendency for the calibration to drift if the sensor is kept at 100% RH for extended periods; furthermore, if the sensor is subject to freezing it can be damaged, particularly if damp. Accuracy is in the order of $\pm 5\%$ RH.

The availability of H^+ ions in the copolymer is dependent on the RH, and this regulates the conductivity of the sensor (Musa & Schnable 1965); interleaving electrodes deposited on its surface allow its resistance to be measured (Fig. 4.6). The resistance change in response to a change in RH is almost logarithmic, ranging from about 1000 ohms at 100% RH to 10 megohms at 5% RH. As with platinum resistance thermometers (PRTs), the resistance of the ion-exchange sensor is measured by making the sensor one arm of a bridge but, as with other sensors that would be polarised by a DC voltage, AC excitation has to be used.

Probably because of the way in which the capacitive sensor has taken over in AWSs, the resistive type is not so widely used. However, development does not stop and the present situation will no doubt change. One area where this type of sensor is used, however, is on the moored buoys operated by the UK Meteorological Office (Fig. 13.2) where it has performed well.

Another form of resistive sensor, no longer much used, was based on a cellulose coating containing small carbon particles in suspension. As the cellulose changed its physical dimensions in response to humidity changes, the contact between the carbon particles changed, varying the resistance.

The dew-point hygrometer

This is a direct and basic method; a mirror is cooled until dew forms on it, the temperature of the mirror then being measured, giving the *dew point* of the air. If the temperature of the air is also known, the VP and RH can be calculated (Wexler 1965). If the dew point is below 0 °C, frost forms instead of dew. Cooling is usually brought about by a Peltier device (this, in effect, is the

Figure 4.6. This ion-exchange sensor is larger than the capacitive sensor of Fig. 4.5; the pin indicates its size. It changes in electrical resistance (rather than capacitance) with changes in relative humidity.

reverse of a thermopile, current passed through it producing heating at one set of junctions and cooling at the other). The dew-point hygrometer is not cheap, owing to its relative complexity. It is, however, a good standard reference technique since it measures dew point directly, all other methods described above relying on the indirect effects of water vapour to achieve the measurement. It is also used to measure very small differences in RH in the Bowen-ratio method of sensing evaporation (Chapter 6).

Dew cells

The propensity of hygroscopic salts such as calcium chloride to absorb water vapour from the atmosphere and become damp can be used to measure humidity. When such a salt is exposed to air containing water vapour, it changes its phase from dry to wet at a particular vapour pressure. By electrically heating a tape soaked in a saturated solution of the salt, the temperature at which the phase change occurs can be measured and the VP derived

(Folland 1975). Because the element cannot be cooled, only heated, the sensor cannot measure a humidity less than that at SVP over lithium chloride at the ambient temperature. At $0°C$ this is equivalent to 15% RH, which is acceptable, while at higher temperatures it is lower still, and so presents no problem. At $-45°C$ the RH is 100% and the instrument is of no use. This type of sensor is now rarely used for environmental applications, in part because it needs to be replenished with lithium chloride regularly but also because it is vulnerable to pollution. It was more commonly used before the development of the thin-film capacitive sensor, which now dominates the field.

Radiation-absorption sensors

Water vapour absorbs infrared radiation, and this phenomenon can be used to measure the amount of water vapour present in the air (Moore *et al.* 1976). Water vapour, and other gases, absorb specific narrow bands over a wide range of electromagnetic wavelengths. Radiation corresponding to any of these bands will lose energy by an amount depending on the amount of water vapour present, the length of the path and the absorption coefficient at the given wavelength. This can be a precise method, but is rarely used for the everyday measurement of humidity because of its complexity and because it consumes power. But the method does have an important application in the measurement of evaporation by the eddy correlation method (Fig. 6.4) (and also the measurement of carbon dioxide fluxes).

Calibration of humidity sensors

The vapour pressure over a saturated salt solution is lower than that over water at any given temperature. This can be used to create atmospheres of known RH for the purpose of calibrating humidity sensors of the thin-film type.

In theory all that needs to be done is to part fill a small container with a saturated solution of a salt such as lithium chloride, place the sensor in the air space above the solution, keep it all at a fixed and known temperature and wait until the humidity stabilises. Equilibrium is reached more quickly if the area of the liquid surface is large compared to the volume of air and if the air is circulated with a fan and the solution is stirred. The temperature must also be kept stable and there should be nothing in the container that might absorb water vapour. Even with such care, it can take several hours for equilibrium to be reached and in the past there was also some uncertainty about the correct RH values for some salts. Tables are available that list the solutions recommended for humidity control (Young 1967) and bottles of saturated solutions can be bought especially for this purpose.

References

Bullot, F. & Dupriez, G. L. (1971) Comparaison d'instruments de mesure de l'humidité sous abri. *Arch. Meteotol. Geophys. Bioklimatol, Wien*, **19B**, 53–6.

Folland, C. K. (1975) The use of the lithium chloride hygrometer (dew-cell) to measure dew-point. *Met. Mag.*, **104**, 52–6.

Jason, A. C. (1965) Some properties and limitations of the aluminium oxide hygrometer. In *Humidity and Moisture, Vol. 1, Principles and Methods of Measuring Humidity in Gases*, ed. A. Wexler, pp. 372–90. Chapman & Hall.

Met. Office (1982a) *The Handbook of Meteorological Instruments*. HMSO, London.

Met. Office (1982b) *The Observer's Handbook*. HMSO, London.

Moore, C. J., McNeil, D. D. & Shuttleworth, W. J. (1976) A review of existing eddy-correlation sensors. Institute of Hydrology Report 32, pp. 22–3.

Musa, R. C. & Schnable, G. L. (1965) Polyelectrolyte electrical resistance humidity elements. In *Humidity and Moisture, Vol. 1, Principles and Methods of Measuring Humidity in Gases*, ed. A. Wexler, pp. 346–57. Chapman & Hall.

Nelson, D. E. & Amdur, E. J. (1965) A relative humidity sensor based on the capacity variations of a plastic film condenser. In *Humidity and Moisture, Vol. 1, Principles and Methods of Measuring Humidity in Gases*, ed. A. Wexler, pp. 597–601. Chapman & Hall.

Salasmaa, E. & Kostamo, P. (1974) New thin film humidity sensor. In *Proc. 3rd Symp. on Meteorological Observations and Instrumentation*, pp. 33–38. American Meteorological Society.

Spilhaus, A. F. (1935) *The transient condition of the human hair hygrometer element*. MIT, Cambridge MA, Meteorological Course, Professional Notes No. 8, Part 1.

Stormbom, L. (1995) Recent advances in capacitive humidity sensors. *Vaisala News* **137**, 15–18.

Wexler, A. (1965) Dew-point hygrometry. In *Humidity and Moisture, Vol. 1, Principles and Methods of Measuring Humidity in Gases*, ed. A. Wexler, pp. 125–215. Chapman & Hall.

Wylie, R. G. (1968) Resumé of knowledge of the properties of the psychrometer. CSIRO, Melbourne, Report No. PIR-64.

Young, J. F. (1967) Humidity control in the laboratory using salt solutions – a review. *J. Appl. Chem.*, **17**, September 1967.

5

Wind

The variable

Wind is caused by imbalances in the atmosphere due to temperature and pressure differences. The movement of air is an attempt to attain equilibrium but, owing to solar heating, this is never achieved. Although air movement is three dimensional, the horizontal component is usually by far the greater and it is this that is normally meant by the term 'wind'. However, vertical motion also occurs, both at a small scale near to the ground, as eddies caused by turbulent flow and convection, and on a large scale as a result of solar heating in the tropics, which powers the general circulation of the atmosphere.

For the first 100 metres or so above the ground, wind speed increases approximately logarithmically (Fig. 5.1), but with increasing height the influence of the surface has progressively less effect and at an altitude of somewhere between 500 and 2000 metres, depending on surface roughness and other factors such as latitude, the speed becomes constant and equal to the *geostrophic wind* (the wind blowing parallel to the isobars). The altitude through which the earth's surface has an influence on the wind is known as the *planetary boundary layer*.

Wind direction is also affected by altitude. At the top of the planetary boundary layer, the direction is the same as that of the geostrophic wind. But descending through the layer, the wind blows at an increasingly oblique angle across the isobars with a component towards the lower pressure region. Plotted from above, the movement of the line of direction marks out a spiral, known as the *Ekman spiral* (Lockwood 1974). While our main concern here is with measuring wind at the surface, a knowledge of what is happening throughout the layer is needed to understand activity at the ground. This is the function of radiosondes (Chapter 13). At the smaller scale, thunderstorms and

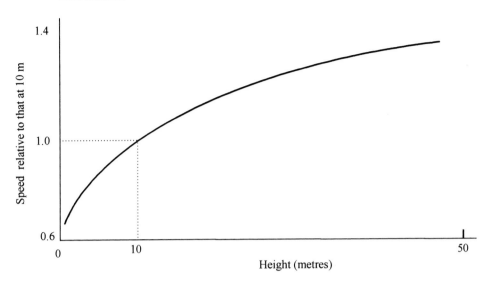

Figure 5.1. Windspeed increases rapidly with height up to about 60 metres, requiring corrections to be made to readings taken at heights other than the standard 10 metres.

tornadoes produce strong local winds with significant vertical components, while in sea- and land-breezes, caused by adjacent parts of the earth's surface being at different temperatures, the warmer air rises and is replaced by the cooler. Mountains also affect the flow of air. The same types of instrument are used to measure winds at all these scales.

Units and terminology

The most appropriate unit today for the measurement of wind speed is metres per second (m s^{-1}) although other units from the past continue in use. As a guide for conversion, 1 metre per second is equivalent to: 2.237 miles per hour, 3.600 kilometres per hour, 3.281 feet per second and 1.943 knots (nautical miles per hour; because nautical miles vary with latitude, the internationally agreed nautical mile is 1852 m or 6076.12 feet, thus 1 knot = 1.15 mph). Tradition also provides a way of estimating wind speed without instruments, by using the Beaufort Scale.

Wind direction is specified as the direction from which the wind is blowing and is expressed in degrees clockwise from true north. The resolution used will vary depending on the application of the data. For most purposes the direction is usually expressed in fives or tens of degrees, or as 8, 16 or 32 compass points.

Exposure of wind sensors

Because wind speed increases rapidly with height, and to allow intercomparison of measurements between sites, 10 metres is used as the standard reference (Fig. 5.1). Nevertheless, many sensors are exposed at a lower height than this, for reasons of convenience or practicality; where 10 metres is not used, it is possible to apply a correction. However, the 10 metre rule only applies to an unobstructed site and if sensors are being operated at anything other than a clear site, they may need to be exposed at a greater height than 10 metres. There are rules for this, which can be abbreviated thus.

Where obstructions are small and surround the site evenly, the sensor is raised 10 metres higher than the obstructions. If the obstruction's distance away is the same as its height, the sensor is raised to about twice that height. If the obstruction's distance away is ten times its height, the sensor is raised to one and a half times that height, and at twenty times to one and a quarter the height. If the obstruction's distance away is thirty times its height, no increase in height is needed. When wind sensors are raised above 10 metres, an *effective height* has to be allotted to them, which would give the same readings as those actually observed by the instrument, taking into account the type, size, number and distribution of the obstacles and the actual height of the sensor. Thus although the sensor may be 25 metres up, its effective height might be judged to be 15 metres. This is a very subjective process and there are no hard-and-fast rules. Whether it is its actual or its effective height that is 15 metres, a correction must be made to convert it to 10 metres. This applies to land situations; over the sea, the corrections are slightly different. If there is no alternative but to instal a wind sensor on top of an isolated building, the procedure is to put it on a mast at least half, and preferably three quarters, the height of the building. This does not apply to lighthouses!

Wind direction measurement

It is possible to make a rough estimate of wind direction by observing the drift of smoke from chimneys, if they are sufficiently far above the surrounding buildings, but care is needed to avoid deceptive perspective. The movement of clouds, even low ones, should not be used because the direction normally changes with height (see the discussion in the first section of this chapter).

Figure 5.2. A manually read wind vane is a simple device, although the mathematics of its movement is less so. Standing beneath the mast, an observer takes an average reading of direction over 10–20 seconds.

Vane aerodynamics

At its simplest, a wind vane consists of a vertical plate on a cross-arm with a counterbalance, the whole rotating at its balance point on a vertical shaft (Fig. 5.2). The vane must have a minimum of friction so that it responds to low wind speeds, it must be balanced at the pivot point, and it must be correctly damped. A compromise has to be reached that allows the vane to reposition itself, after a change of wind direction, without excessive delay, but the damping must not be so minimal that the vane overshoots the new position (Wieringa 1967). This characteristic is quantified as the vane's *damping ratio*, which is the ratio of the actual damping of the vane compared with its *critical damping* – the value of damping that would give the fastest movement to the new direction without any overshoot. In practice it is best for this ratio to lie between 0.2 and 0.7. The ratio is also known as the *damping coefficient*. It is a constant for any vane, being independent of other variables such as wind speed or air density. Another related term is the *damped wavelength*, which is the length of the column of air that passes the vane during the time it takes to move through one damped oscillation. This too is a constant for any vane. A similar term, the *delay distance*, is the length of the column of air that passes while the vane responds to 50% of a step change in direction. The *distance constant* (which is like a time constant) is the length of the column of air that passes while a vane responds to 63.2% of a change in direction, a typical value for this being about four metres.

The damping will depend on such factors as the size of the vane, its inertia and the natural aerodynamic damping that occurs when a plate is moved through a gas. While friction can be introduced to increase damping, it has the disadvantage that its effect reduces with increasing speed, making the vane behave poorly at higher speeds.

The majority of vanes have a single plate. A few have two – either flat, some distance apart and parallel to each other, or as two parallel aerofoils. Another alternative is two flat plates splayed out into a wedge shape from the shaft. According to Wieringa (1967) the wedged-shaped vanes are likely to reduce each other's effectiveness, while the streamlining of the aerofoil plates may reduce their effectiveness at low speeds.

Manual and remote-indicating wind vanes

The simplest type of wind vane is shown in Fig. 5.2. An observer looks at it from beneath and estimates the average direction over about 15 seconds.

In the pre-microelectronic era, if the direction were to be displayed remotely, a Desynn or Magslip transmitter would be attached to the vane's shaft with a cable connecting it to the appropriate receiver in a nearby office, the dials of the receivers being marked in 5 or 10 degree steps along with the main compass points. The receivers contained a magnetic rotor that took up the same angle as the transmitter rotor, the transmission of the information requiring just three wires. A comparison of the two types of transmitter is not relevant since today a shaft encoder would more likely be used instead (see the next subsection). It is possible to make the receiving Magslip or Desynn machine move a pen across a paper chart, and so record the direction.

Automatic wind direction sensors

The great majority of wind direction sensors for automatic logging use a vane, the angular position of the shaft being sensed electrically. Other methods than vanes are used, but these will be dealt with in the section on wind speed under the headings of pressure, thermal and sonic anemometers.

Sensing shaft angle by potentiometer   The angular position of a shaft can be measured electrically either by using a *potentiometer* or a shaft encoder. Potentiometers can be of the wire-wound or conductive plastic type or they can be constructed as a circular array of magnetic reed switches connected to a resistor network (Fig. 5.3). By using a diode encoder with the reed assembly, a binary signal similar to the Gray code (described below for shaft encoders) can

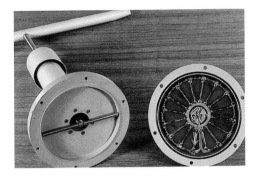

Figure 5.3. This wind direction sensor uses reed switches to convert the shaft rotation into an analogue voltage signal, the magnet arm (left) closing at least one of the 16 reed switches (right), spaced at 22.5 degrees. The reed switches tap off a voltage proportional to direction from a series of fixed resistors. In some designs (Fig. 5.6) a compact, wire-wound potentiometer is used in place of reed switches.

be generated instead of the analogue voltage signal produced by potentiometers.

A problem with wire-wound, and to a lesser extent plastic, potentiometers is that they become worn, particularly in the direction of the prevailing wind, because the vane moves the wiper contact backwards and forwards over the same part of the track continuously. Reed switches avoid this problem while also reducing the torque needed to move the vane; this can be high, reducing the vane's sensitivity at low wind speeds. A typical starting speed quoted by manufacturers, its *threshold*, is about $0.3 \, \text{m s}^{-1}$, although it can be as high as $1 \, \text{m s}^{-1}$. (But threshold information is only meaningful if the angle of attack of the wind is also quoted, and it rarely is.)

The resolution of a wire-wound potentiometer is high, limited only by the closeness of the wire turnings, and for a plastic potentiometer it is theoretically infinite. The relative coarseness of resolution of an array of reed switches is offset by the fact that usually the wind direction is constantly changing through many degrees and so an average over a period of minutes removes much of the resolution problem. In addition, wind direction is rarely needed to high precision. Where high resolution of spot readings is needed, a wire-wound or plastic potentiometer, or a shaft encoder, will be required.

When the wind direction is alternately left and right of north, the output of the potentiometer changes from high to low. If an arithmetic average of several such spot readings is taken, the resultant indication can be south instead of north, although if individual spot readings are recorded the problem does not arise. The simplest solution is to take vectorial means rather than arithmetic means. But an alternative strategy is to extend the angular range of the sensor artificially so that, for example, it works over a range of ± 270 degrees, a logic circuit connected to the reed switch assembly switching to a stable position whenever the ends of the range are approached, thereby avoiding any step change.

Digital shaft encoders   An alternative to potentiometers for measuring angular position is the shaft encoder. The most common has an optical disc encoded with a binary pattern, read using infrared light-emitting diodes (LEDs) on one side of the disc and a photodetector on the other, producing a binary code (often, the Gray code; see binary numbers, Chapter 11). Depending on the angular resolution required of the sensor, the disc may be encoded with as few as four tracks (giving a 4-bit word) or as many as nine for high resolution. More information on shaft encoders is given in Chapter 10 under the topic of water level measurement.

Damping ratios   Damping ratios for most of the automatic wind direction sensors vary from 0.2 to 0.4. Values of the *damped wavelength* typically range from 2.5 to 7 metres.

Measuring wind speed

The Beaufort Scale

In 1806 Admiral Sir Francis Beaufort devised his Beaufort Scale for wind speed. The scale subdivides wind into strengths on a scale of 0 to 12, each defined by their observable effects. There are two sets of descriptions, one for land and the other for sea (Met. Office 1995, see the Chapter 1 reference list). Just the land version is summarised in Table 5.1, in abbreviated form. Comparisons between the above force numbers B and measurements of wind speed V at 10 metres have led to the relationship $V = 0.836\sqrt{B^3}$, where V is in metres per second.

Cup and propeller anemometers

Sensing wind speed with cups   The cup anemometer was invented in 1846 by Dr Thomas Romney Robinson, a clergyman and astronomer from Armagh (Patterson 1926). The design has changed little since then. When the anemometer is exposed to the wind, the pressure on the open side of the cups is greater than that on their backs, which causes the shaft to rotate (Fig. 5.4). The speed of rotation is nearly linear with respect to wind speed and only for more precise

Table 5.1. *The Beaufort Scale*

Name	Description	Force, B	Mean wind speed V	
			knots	m s^{-1}
Calm	Smoke rises vertically	0	0	0
Light air	Smoke drifts	1	2	0.8
Light breeze	Wind felt on face	2	5	2.4
Gentle breeze	Light flag extends	3	9	4.3
Moderate breeze	Dust raised	4	13	6.7
Fresh breeze	Small trees sway	5	19	9.3
Strong breeze	Umbrellas hard to use	6	24	12.3
Near gale	Whole trees move	7	30	15.5
Gale	Twigs break off trees	8	37	18.9
Strong gale	Slight structural damage	9	44	22.6
Storm	Trees uprooted	10	52	26.4
Violent storm	Widespread damage	11	60	30.4
Hurricane	No description	12	—	—

work is any correction required. The response is independent of wind direction and, more or less, of air density. Robinson assumed that the ratio of the speed of the wind to the speed of the cup-centres was a constant, and equal to 3. But this ratio – the *anemometer factor* – in fact depends on wind speed and instrument dimensions (Patterson 1926), and has subsequently been shown to be yet more complex (Ramachandran 1970). It is not possible to determine theoretically the best design for a cup anemometer, and improvements have come about through wind-tunnel tests, in which each of the cup characteristics is changed individually.

From experiments on a range of anemometers with different cup diameters and shapes and various arm lengths (Acheson 1970, Hyson 1972) it was found that three cups are better than four, because the torque is more constant over a complete revolution and because three cups give more torque weight-for-weight. It was also found that semiconical cups performed better than hemispheres and that beaded edges to the cups rather than a plain finish made the instrument less sensitive to turbulence.

MacCready & Jex (1964) showed that the time constant of a cup anemometer varies inversely with the wind speed, and so it is more meaningful to express the time response of the system in terms of the length of the column of air that passes the anemometer while it responds to 63.2% of a step change in wind speed, this being called the *distance constant* (see earlier in the chapter); it is the product of the time constant and the wind speed. Because the time constant varies with wind speed, the cups speed up more quickly with an increase of wind speed than they slow down with a decrease. The effect of this is

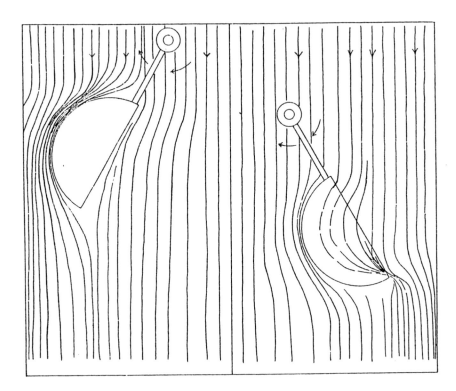

Figure 5.4. The flow of wind around a cup anemometer produces a higher pressure on the open side of the cups than on their backs, giving rise to a torque that rotates the shaft (Patterson 1926).

that in a variable wind the mean speed measured by the anemometer is higher than it should be.

Schrenk (1929) derived a theoretical relationship expressing the extent of overrun in varying winds and found that the larger the cups, the longer the arms, the slower the changes in wind speed, the higher the mean wind speed and the smaller the mass of the moving parts, the smaller will be the overrun error. As the gustiness of the wind increases, the greater are the errors, so that when the wind fluctuations are large compared with the mean speed, the overrun can be as much as 30% for anemometers with design characteristics that are the exact opposite to those listed above; nevertheless, Hyson (1972) believed that overrun generally amounts to about only 1% in the real world and that anything in excess of this is probably due to another source of error – the vertical wind.

Sensing wind speed with propellers   The next-most-common sensor of wind speed is the propeller or fan anemometer, described by Dines in 1887. It has three or four blades in the form of helicoids rotating on a horizontal shaft. To keep the propeller facing into the wind, it is usually combined with a wind-direction sensor vane. What has been said about cup-anemometer response applies equally to the propeller, although being more linear it is more precise. There is the extra consideration, however, of the *yaw angle*, since the propeller will not always face directly into the wind owing to the time it takes to respond to a change of direction. Mazzarella (1954) made measurements that suggested that the ratio of measured to actual speed was $\cos^2 \alpha$, where α is the yaw angle. However, Monna & Driedonks (1979) suggested $\cos(1.3\alpha)$, although this value may be dependent on wind speed.

In the case of the combined propeller-and-vane construction, the vane performance may be affected by the propeller in a way that is not easy to quantify, and the gyroscopic effect of the propeller rotation may also produce an increase in the vane's damping ratio.

Because propellers are directionally sensitive, they also find an application in situations where it is necessary to measure the three-dimensional nature of air movement, such as in turbulence studies. Such an instrument consists of three propellers, mounted at right angles on a common mast, measuring the along-wind, across-wind and vertical-wind components. Each propeller measures the wind speed parallel to its axis, the response to wind at an angle approximating to a cosine law although the response slows and the distance constant increases with increasing angle. As well as the speed, the direction of rotation of the propellers also has to be sensed.

Measuring cup rotation mechanically   The most usual way of measuring the rotation of a cup anemometer shaft for making manual observations is with a mechanical revolution counter (Fig. 5.5), readings being taken at set times (three-hourly or daily, for example), the difference between the readings giving the amount of *wind run* over the period (easily converted to average speed). This is not a suitable method for obtaining spot readings of instantaneous speed; for these, hand-held cup anemometers are available, with mechanical or electronic displays, but they are not meant for precise work.

Measuring rotation by generator   For spot readings, the shaft may drive a permanent-magnet AC generator, producing a voltage proportional to speed, the voltage being displayed on a meter some distance away. The AC voltage can also be converted to a DC signal and logged at automatic stations as spot readings. An alternative is to measure the frequency of the AC signal rather

Figure 5.5. A manually read cup anemometer totals the number of revol-utions of the shaft over a period, each turn being proportional to the passage of a known 'run of wind'. Given the elapsed time between readings of the counter, the total run of wind (or the mean wind speed) can be calculated.

than the voltage, giving either a spot reading or (by integration of the cycles) the run of wind. The starting torque of the larger generator anemometers can be high because the magnetic attraction between rotor and stator offers resistance to movement. Generators are used more widely with propeller anemometers than with cups.

Measuring rotation by switch   For the automatic logging of wind speed, cup anemometers that produce a switch contact closure are generally the most useful (Fig. 5.6), the switch being either a magnetic reed switch or an optical equivalent. In most designs, the rotating shaft has a magnet fixed to it that closes a switch once for each revolution. The closures are processed by a data logger, their total over a period being logged to give the run of wind or the average speed, with the possibility also of measuring the time between pulses so as to log instantaneous speeds or peak gusts.

In the pre-microelectronics era, a pulse for each revolution of the cups was too much to handle. To overcome this, a worm-reduction gear was included that caused a magnet to fall past the reed switch for the passage of every tenth of a kilometre (or of a mile) run of wind.

Reed switch operating lifetimes are around 10 million closures, but eventual-ly they fail. An alternative is to use an optical disc through which an LED passes infrared (IR) light, which is then detected and produces one pulse per

Figure 5.6. An electrical anemometer. For automatic measurements, the revolutions of the cups are usually detected by a reed switch closed at each rotation by a magnet, as in this example, although optical sensors are also used. The inverted direction sensor measures shaft angle with a small continuous-rotation potentiometer.

revolution. Alternatively, the disc may be encoded with an optical pattern that produces many pulses per revolution, which can be treated as a frequency or as individual pulses, giving an indication either of instantaneous speed or of total counts over a period. A slight disadvantage of the optoelectronic method is that it takes power, while reed switches require none, although consumption can be kept low enough to be acceptable.

Cup anemometers vary widely in size and shape, depending on their intended use. Some of the large and heavier designs, meant for general use over long periods, may not start to turn until wind speeds of around 0.5 to 1.0 m s^{-1} are reached, while the more sensitive can respond to speeds down to 0.1 m s^{-1}. Top speeds of between 50 and 75 m s^{-1} are usual. Most manufacturers do not quote a distance constant, but where they do it falls between one and three metres.

Pressure anemometers

Although cup and propeller anemometers dominate the field, other ways of measuring wind have been developed, such as sensing its pressure.

Pressure-tube anemometers   Henry De Pitot, a student of Reaumur in Paris, was superintendent of the Canal du Midi in Languedoc. His greatest claim to fame rests on his invention of the *Pitot tube* in 1732, which he developed for the measurement of the flow of rivers, rather than air. He was mistaken over the theory of its operation but it was nevertheless a useful invention that helped answer questions about the velocity of flow in rivers at different depths, then an unknown factor (Chapter 10).

The Swiss family of Bernoulli (Johann, Jacob and Daniel) were all mathematicians and somewhat competitive amongst themselves. It was Daniel Bernoulli, in 1738, who analysed pressure–velocity relationships and, although he was unable to derive a general expression, he solved some special cases; the equation

$$p_t = p_s + 0.5\rho u^2$$

is now generally known as the Bernoulli equation. Here the *total head* p_t is made up of two parts, the *static head* p_s and the *speed head* $0.5\rho u^2$, where ρ is the density, and u the speed of the fluid. Although derived for the flow of water the same principle also applies to the flow of air and so to the measurement of wind speed, the difference between the two pressures being measured using a Pitot tube (Fig. 5.7). The tube is kept facing into the wind, giving the total head, while the holes in the wall of the tube give the static head.

Dines (1892) wrote a paper comparing various types of anemometer, including the Pitot tube. His own design, now obsolete, differed slightly from the modern version described above in that the holes that measured the static head were in the vertical tube. This had the effect of causing some suction, thereby reducing the static pressure slightly. Dines describes a method of recording the pressure difference, transmitted through tubes from the sensor head, using a float in a tank of water with pulleys and threads to move a pen over a chart. He concluded that the measurement was very uncritical of alignment with the wind. In fact even with a misalignment of 15° or more, the error in speed indication was less than 1%.

Pitot tubes are not much used today, although it would be possible to measure the pressure difference quite easily with a modern differential pressure sensor. However, a specialised design of pressure anemometer has been developed that measures the pressure on the surface of a tube held vertically in the air-stream through holes spaced at 90° around it, pointing at the four compass points. The tube is divided internally into four chambers behind the holes; differential pressure sensors are connected between the N and S chambers and between the E and W chambers, giving voltage outputs that vary in amplitude and polarity, from which the speed and direction of the wind can be calculated.

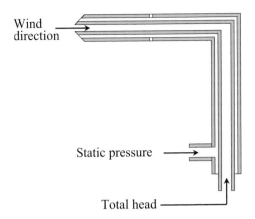

Wind direction

Static pressure

Total head

Figure 5.7. In a Pitot tube sensor, the difference in pressure between the total head and the static head is proportional to the wind speed. The holes sensing the static pressure are in the horizontal tube in a modern design, although in an earlier design by Dines they were in the vertical tube, introducing some slight error.

An accuracy varying from $\pm$ 0.5 to $\pm$ 2.5 knots over the range 0 to 150 knots is claimed by the manufacturer. There is the dual advantage that the tube can be heated to prevent ice formation and does not have to be kept pointing into the wind.

Drag-force anemometers   When a rotating sensor is impractical, drag or thrust anemometers can be a useful alternative and several have been developed (Norwood *et al.* 1966). They can also be useful where the sensing of high-frequency wind fluctuations is important or where it is the dynamic pressure that is the primary concern (as opposed to wind speed). They furthermore have an application where three-dimensional wind measurements are required (Doe 1967, Smith 1970), serving much the same purpose as the three-propeller system described earlier, but with a faster response.

The earliest form of anemometer was a plate, kept at right angles to the wind and free to swing about a horizontal axis, the angular deflection due to the pressure of the wind being a function of wind speed. The *normal-plate anemometer* is a development of this, consisting of a plate held stationary and perpendicular to the wind. Dines (1892) describes such an instrument, comparing it with a Pitot tube, cup and propeller anemometer, and shows how its readings can be recorded using a float in water similar to the one he describes for the Pitot tube.

The drag force on a body is given by $F = \frac{1}{2}A\rho u^2 C_d$ where F is the drag force, A the area of the body, ρ the density of the air, u the air speed and C_d the drag coefficient of the body (which is dimensionless, and typically around 0.7). Drag anemometers are usually in the form of a cylinder for two-dimensional measurements (not being badly affected by vertical components in the wind) or a sphere for three-dimensional wind studies.

In the 1880s Osborne Reynolds developed his theories of fluid flow, forming the basis of present-day aerodynamics (Kuethe & Schetzer 1950). Part of Reynold's work concerned *bluff bodies* (cylinders and spheres), for which the major cause of fluid drag is 'form drag', whereas for a streamlined object the main cause of drag is skin friction. He found that as the speed, u, of the fluid (in this case air) increased, changes occurred in the nature of the flow, depending on the size of the object (in the case of a cylinder its diameter, d), on the viscosity of the fluid, μ, and on its density, ρ, the ratio $\rho u d / \mu$ being known as the Reynolds number, R_e.

Above a certain critical value of R_e, the flow switches from smooth to turbulent, going through various intermediate stages as the speed increases. Up to an R_e of 3, the flow is entirely smooth; from $R_e = 3$ to 40, vortices form in the shadow of the cylinder, but the flow reforms downstream; above an R_e of 40 it does not reform, while above about 70, a train of eddies, known as a *Karman vortex street*, develops (Fig. 5.8). Above $R_e = 200$ the vortices break up into smaller regions of vorticity, the street reforming downstream. When R_e reaches about 10^5, the stresses become so great that turbulence occurs even in the *laminar boundary layer* adhering to the bluff object, whereupon the drag force suddenly drops. These changes are of significance in the design of a drag anemometer. For example, for a cylinder with a diameter of 8 cm the speed at which the drag force suddenly drops is $50 \, \text{m s}^{-1}$; a larger cylinder might, therefore, mis-measure higher speeds. While the drag coefficient is not constant for all values of R_e, for an 8 cm cylinder it is relatively so for R_e values of 100 to 10^5, covering most wind speeds.

Instruments of this type usually sense the pressure of the wind on a cylinder or sphere, held stationary, by means of strain gauges fixed to the structure on which the cylinder or sphere is mounted (Fig. 5.9). Other methods have been used in which the drag body moves slightly against a spring, the magnetic coupling between a ferrite core and four coils being used to detect the movement. To prevent oscillation, oil damping may be used. An advantage of drag anemometers is that both speed and direction can be measured by the one instrument. Also, by heating the drag body it is possible to deter the formation of rime ice or the adherence of wet snow, although in one design ice was removed pneumatically (Chapter 13).

However, drag anemometers have many problems – the zero can drift and because of their square-law response, their accuracy varies greatly across the

Figure 5.8. Eddies form what is known as a Karman vortex street down-stream of a body exposed to the wind, their frequency giving a measure of wind speed. In the illustration, they are forming in the wake of a cylinder at a Reynolds number of about 70. The arrow shows the direction of the wind.

Figure 5.9. A drag-force anemometer senses the force of the wind as it passes round a fixed shape, in this case a cylinder, here temporarily shielded from the wind by a cover, allowing the zero setting to be checked. It is part of the cold regions AWS shown in Fig. 13.9.

speed range, being poor at low speeds if a wide velocity range has to be measured. They can also be affected by rain.

Vortex anemometers

Another somewhat specialised instrument uses the fact that in the lee of an obstruction, such as a cylinder, Karman vortex streets form (see above, and Fig. 5.8). These vortices are not shed randomly but at a very specific frequency, which Strouhal defined as $f = u/Sd$, S being the Strouhal number, u the velocity of flow, d the diameter of the cylinder and f the frequency. S is constant, at about 5, throughout all the Karman-vortex-street range of Reynolds numbers. So, for example, at a speed of $50\,\mathrm{m\,s^{-1}}$ and with a cylinder of diameter 1 cm the frequency is 1000 Hz, while at 5 m s^{-1} it is 100 Hz. The spacing of the vortices is about 2.5 times the diameter of the cylinder. An advantage of the method is that the frequency is linearly related to speed and so (theoretically) no calibration is required. Accuracy in operation is simply a matter of ensuring that the vortices are reliably detected. There can be no zero drift or calibration change.

The vortices can be detected by passing an ultrasonic sound wave through their path, the speed and direction variations within the vortices modulating the beam; an alternative method of detection is the use of a *hot wire sensor*. The construction of a vortex anemometer is very much like that of a propeller anemometer; it has a vane to sense the wind direction, and the vortex-producing cylinder, with its down-wind vortex sensor, is mounted on the vane's arm to keep it pointing into the wind.

Thermal anemometers

There are three main types of sensor that use heat to measure wind speed, and, in some cases, direction also. These are *heat-transport*, *heat-pulse* and *heat-loss* anemometers, the latter being the most common although none have wide use; they are all designed for situations in specialised micro-meteorological research where a rapid response is needed. All methods involve heating a small resistance wire by passing a current through it and measuring the effects caused by the wind.

Heat-transport anemometers   A heater wire, powered continuously, produces a stream of heated air which is carried downwind (Dyer 1960, Taylor 1958). Mounted each side of the heater wire, and just a few millimetres away, are heat-sensing detectors, usually in the form of fine platinum wire resistance thermometers. By mounting the detector wires at right angles to the heater wire, the heater being vertical, a wide yaw response is achieved. Wind speed is measured by detecting the difference in temperature between the detector wires resulting from the transport of heat (advection) in the direction of the wind. These instruments have a cosine response over a wide range of angles and can indicate from which direction the wind vector is oriented. However, calibration is difficult at low speeds and, as with all wind sensors using heat, they can be affected by rain and solar heating, although with screening this can be avoided.

Heat-pulse anemometers   The physical design of this sensor is similar to that of the heat-transport anemometer, but the heater wire is pulsed to a high temperature by the passage of current for a few microseconds, introducing a tracer of heated air, the time of flight to the detector wire giving the wind velocity. The method has the advantage of being sensitive to very low wind speeds, although its upper limit is restricted to about $15 \, \mathrm{m \, s^{-1}}$. It has, however, a very fast response, with a short distance constant that is dependent on the distance between the wires, which can be made very small; because it measures the time for the pulse to travel a known distance its output is directly related to wind speed. Wind direction can also be inferred, but the electronics is complex.

Heat-loss anemometers   In the previous two cases, heat transported down-wind was detected to give a measure of wind speed. In heat-loss instruments, the amount of heat lost from a heated wire is measured.

In the *hot-wire anemometer*, a short length of platinum resistance wire is supported vertically in the wind and heated by passing a current through it, the rate of heat loss being measured. The loss is made up of two components – that which occurs in still air due to radiation and convection from the wire and that due to the advection of heat when a wind moves past the wire.

The theory of this process was analysed by King (1914), who showed that if the resistance wire was heated to a constant temperature (and thus its resistance was also constant), $I^2 = I_o^2 + Bu^{1/2}$ where I is the total current, I_o is the current in still air, B is a constant determined by experiment and u is the wind speed. The majority of hot-wire anemometers operate in this constant temperature mode (Perry & Morrison 1971) because it gives a better response at higher wind speeds than the alternative, simpler, constant-current method, in which the temperature of the wire varies with wind speed (Miyake & Badgley 1967).

If platinum wire with a diameter of 0.002 mm is used, the time constant of the system is very small and the sensor can follow changes of speed down to hundredths of a second. However, this results in a rather delicate instrument and for more rugged designs thicker wires or *hot films* are used, made by depositing a thin layer of platinum onto an insulating support, such as a cylinder a millimetre or so in diameter.

Heat-loss anemometers are affected by the ambient air temperature unless they are operated well above it. For this reason temperatures as high as $900 \, ^\circ\mathrm{C}$ may sometimes be used, but this can be a problem at low wind speeds because there can be excessive free convection. The simpler sensors do not detect the direction of the wind, although by operating two horizontally deployed wires at right angles, the direction can be calculated (Gjessing *et al.* 1969). The Viking Mars Lander used a multiple hot-film sensor for measuring the speed and direction of Martian winds (Hess *et al.* 1972). Some designs allow three-dimensional measurements of wind speed. Unlike the heat-transport and heat-pulse sensors, the hot-wire anemometer does not have a linear response to wind speed, as the above equation, derived by King, shows.

The great majority of thermal anemometers are used for research into turbulence, only a few finding applications where cups might normally be expected instead. The small number available commercially have an accuracy of about $\pm 5\%$ of the actual speed and $\pm 3°\text{--}4°$ in the direction. However, hand-held hot-wire sensors with a portable meter are more common and these can be particularly useful for measuring low wind speeds in small spaces where portable cup anemometers may not be practicable. A typical specification for this type of instrument ranges from $\pm 3\%$ at the high end of the speed range to $\pm 10\%$ at the lower end.

Sonic anemometers

The speed of sound in moving air is its speed in still air plus the speed of the air. By measuring the difference in time it takes sound to travel with the wind and against it, its speed can therefore be calculated. If the wind is moving at an angle to the sound wave, it is the wind speed component in that direction that is measured, but by keeping the sound path directed in line with the wind (as with a propeller or vortex anemometer) the actual wind speed and direction can be measured. In practice, though, this is usually achieved by using two fixed sound paths at right angles. (The temperature also needs to be measured because the speed of sound is temperature dependent.) It can be shown that

$$u = [c^2/(2l)] (t_2 - t_1),$$

where u is the speed of the wind, c the speed of sound, l the path length and t_1 and t_2 the times for the sound to travel with and against the wind.

The sound can be either a continuous wave or a short pulse. In the former, the difference in phase of the two sound waves is measured while in the latter it is the time of flight of the leading edge that is detected. In early designs, continuous waves were used (Kaimal & Businger 1963), but reflections and extraneous noise from outside caused problems, as did the precise detection of the phase difference, especially near zero wind speed. However, the pulse method requires precise detection of the leading edge (Mitsuta 1966) and this too has its problems, although currently it is the most-used method. Three sound paths are often included, allowing wind speed and direction to be determined in three dimensions (Fig. 5.10).

The technique has a good cosine (i.e. angular) response and can work over a speed range of 0.02 to 40 m s^{-1}. Because the measurements are made over a path length, as distinct from the single-point readings of all the other methods,

Figure 5.10. This ultrasonic anemometer consists of three pairs of transducers, the ones at the top facing those diagonally opposite at the bottom.
Each pair acts alternately as a transmitter and receiver of ultrasonic pulses of
sound, the time of flight between each pair giving a measure of the sound in
the direction of the line between them and, thus by vector calculation, of the
velocity of the wind in three dimensions.

the result is an average of the speed and direction within the sound path. Rain
can cause some loss of data, although the sloping face of the lower sensors (Fig.
5.10) minimises this, and at certain wind directions the physical presence of the
sensors produces a wake that affects the readings. Accuracies for speeds
averaged over ten seconds are in the region of $\pm 3\%$ to $\pm 5\%$ of full scale,
depending on the speed, and of $\pm 2°$ to $\pm 4°$ for direction. They have the
advantage that there are no moving parts to wear out, but their high cost, and
the need for power to operate them, go against their general use, although this
may be beginning to change. However, they are essential tools for research into
turbulence and are useful in evaporation measurement.

References

Acheson, T. A. (1970) Response of cup and propeller rotors and wind direction vanes
to turbulent wind fields. *Met. Monographs*, **11**, No. 33, 252–61.

Dines, W. H. (1892) Anemometer comparisons. *Quart. J. Roy. Met. Soc.*, **17**, 165–85.

Doe, L. A. E. (1967) A series of three-component thrust anemometers. In *Proc. Int. Canadian Conference of Micromet., Part 1*, pp. 105–14.

Dyer, A. J. (1960) Heat transport anemometer of high stability. *J. Sci. Inst.*, **37**, 166–9.

Gjessing, D. T., Lanes, T. & Tangerud, A. (1969) A hot wire anemometer for the measurement of the three orthogonal components of wind velocity, and also directly the wind direction, employing no moving parts. *J. Sci. Inst.*, **2**, Series 2, 51–4.

Hess, S. L., Henry, R. M., Kuettner, J., Leovy, C. B. & Ryan, J. A. (1972) Meteorology experiments: the Viking Mars Lander, *Icarus* **16**, 196–204.

Hyson, P. (1972) Cup anemometer response to fluctuating wind speeds. *J. Appl. Met.*, **11**, 843–8.

Kaimal, J. C. & Businger, J. A. (1963) A continuous wave sonic anemometer thermometer. *J. Appl. Met.*, **2**, 156–64.

King, L. M. (1914) On the convection of heat from small cylinders in a stream of fluid: determination of the convection constants of small platinum wires with application to hot-wire anemometry. *Phil. Trans. A*, **214**, 373–432.

Kuethe, A. M. & Schetzer, J. D. (1950) *Foundations of Aerodynamics*. John Wiley & Sons, London.

Lockwood, J. G. (1974) *World Climatology: An Environmental Approach*, p. 330. Edward Arnold, London.

MacCready, P. B. & Jex, H. R. (1964) Response characteristics and meteorological utilisation of propeller and vane wind sensors. *J. Appl. Met.*, **3**, 182–93.

Mazzarella, D. A. (1954) Wind tunnel tests on seven aerovanes. *Rev. Sci. Instrum.*, **55**, 63–8.

Mitsuta, Y. (1966) Sonic anemometer-thermometer for general use. *J. Met. Soc. Japan*, **44**, 12–24.

Miyake, M. & Badgley, F. I. (1967) A constant temperature wind component meter and its performance characteristics. *J. Appl. Met.*, **6**, 186–94.

Monna, W. A. A. & Driedonks, A. G. M. (1979) Experimental data on the dynamic properties of several propeller vanes. *J. Appl. Met.*, **18**, 699–702.

Norwood, M. H., Cariffe, A. E. & Olszewski, V. E. (1966) Drag force solid state anemometer and vane. *J. Appl. Met.*, December 1966.

Patterson, J. (1926) The cup anemometer. In *Proc. Royal Soc. of Canada*, Series III, Vol. XX, Meeting of May 1926, pp. 1–54.

Perry, A. E. & Morrison, G. L. (1971) A study of the constant-temperature hot-wire anemometer. *J. Fluid Mech.*, **47**, 577–99.

Ramachandran, S. (1970) A theoretical study of cup and vane anemometers, Part II. *Quart. J. Roy. Met. Soc.*, **96**, 115–23.

Schrenk, O. (1929) Errors due to inertia with cup anemometers in fluctuating winds. *N.Z. Tech. Phys.*, **10**, 57–77.

Smith, S. D. (1970) Thrust anemometer measurements of wind turbulence, Reynold's stress and drag coefficient over the sea. *J. Geophys. Res.*, **75**, 6758–70.

Taylor, R. J. (1958) A linear unidirectional anemometer of rapid response. *J. Sci. Inst.*, **35**, 47–52.

Wieringa, J. (1967) Evaluation and design of wind vanes. *J. Appl. Met.*, **6**, 1112–14.

6
Evaporation

The variable

The rate of evaporation is controlled by the relative humidity and temperature of the air, the amount of net radiation, the wind speed at the surface, the amount of water available, the nature of the surface (for example its roughness) and the type of vegetation. Open water presents another situation, as do ice and snow. The net incoming energy is apportioned to the three fluxes – sensible, latent and soil heat – according to the infinite variety and combination of circumstances.

It is much more difficult to sense the rate of loss of water from the surface through evaporation – the latent heat flux – than it is to measure the gain of water through precipitation. Nevertheless, the rate of evaporation is expressed in the same units as precipitation: it is the equivalent depth of liquid water lost into the atmosphere as water vapour, expressed in millimetres lost over an hour or a day.

Measuring and estimating evaporation

Evaporation can be measured either as the loss of liquid water from the surface or as the gain of water vapour by the atmosphere, but few of the methods involve a direct measurement, most inferring the amount by indirect measurement.

Measuring the loss of liquid water

Measuring water loss by the use of a natural catchment   The water-balance equation $E = P - (V_r + V_s + V_1)/A$ relates the various elements in a catchment balance, E being the evapotranspiration, P the precipitation, V_r the runoff in rivers, both above and below the surface, V_s the volume of water stored in the system, V_1 the loss due to leakage and A the area of the catchment.

By measuring the precipitation, the runoff in rivers and the amount of water stored in the ground (the soil moisture), the losses (evapotranspiration and leakage) can be calculated (Rodda *et al.* 1976). The measurement of rainfall, of river flow and of soil moisture (see future chapters) can be done to varying degrees of precision. Provided, then, that the catchment is a simple one, with bedrock beneath that is known not to leak and with no other ways out for the water, so that V_1 is zero, the method is workable, although expensive.

However, the error will be large unless the variables are measured to high accuracy, the evapotranspiration E being the small difference between the three much larger values P, V_r, V_s. Any error in one of these will exaggerate or minimise the apparent evaporation, perhaps by many times its real value. Also the equation only tells us what the evaporation is from the whole catchment, lumping together evaporation from the ground and plants and transpiration. Nor does it explain any of the details of the evaporation process. It can, however, answer such questions as 'Do forests evaporate more water than grassland?' (Law 1956).

Measuring evaporation through soil moisture changes   Provided there is little drainage from the soil (either on the surface or below), the measurement of changes in soil moisture content can be attributed entirely to evapotranspiration, with allowances for rainfall. Accuracy is improved if measurements are also made of *soil tension* (Chapter 9), from which the vertical motion of the soil moisture can be established, helping to quantify any downward movement to the water table that does not contribute to evaporation. However, it is difficult to select a representative piece of land to monitor, since evaporation and soil moisture can vary a great deal spatially. Furthermore, the method does not take account of the evaporation of rain directly from the surface of plants, especially large ones; this water never gets to the soil. The method is, therefore, limited in use.

Measuring water loss using a lysimeter   A lysimeter is a container that is isolated from the surrounding soil but filled with soil similar to that outside it and planted with the same vegetation that surrounds it. By suitable design, the contents of the lysimeter can be made to behave in a similar way to their surroundings, provided that the temperature of the soil is kept the same and the drainage is similar (Fig. 6.1). To produce the correct drainage it may be necessary to instal some means of suction at the base of the container, to ensure that the soil moisture tension is the same as in the freely draining ground

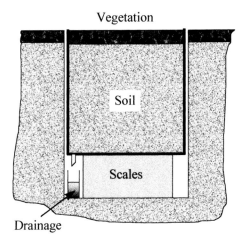

Figure 6.1. A lysimeter is a container filled with soil, replicating as closely as possible that outside, and with the same vegetation. It is weighed either with a simple flexible container filled with a fluid with a manometer tube to the surface or with a mechanical or electronic balance. Changes in weight indicate both precipitation and evaporation. Excess water is drained at the base and measured.

outside the lysimeter. 'The bigger the better' is generally the case for lysimeters, since then edge effects are reduced and internal and surface differences are smoothed out. However, small lysimeters can still give very useful results.

There are two main types of lysimeter – those that measure the weight of the container and its contents and those that do not. John Dalton experimented with the measurement of evaporation in 1795 at a site near Manchester, using what we would now call a non-weighing lysimeter (Biswas 1970). It took the form of a container 25.4 cm in diameter and 1 m deep, with a tube at the bottom to collect drainage and one at the top to collect surface runoff. It was filled with sand and gravel at the bottom, with soil on top, left bare for a year and then covered with grass for the second year. A record was kept of the rainfall and the quantity of water collected from the two tubes for two years. Dalton concluded that the annual evaporation was 25 inches (63 cm) and that evaporation increased with rainfall (but not proportionally) and was the same for soil and grass.

In the case of a weighing lysimeter, a mechanical balance is the best way of measuring the weight because it is the very small changes in weight (due to rainfall and evaporation), not the absolute total dead weight of soil and container, that are wanted. Load cells can be used, but they weigh the total mass and so must be able to resolve small changes in weight in excess of the very large overall weight. It is also possible to use a hydraulic method of weighing, in which the lysimeter stands on a flexible container filled with oil connected by tubing to a manometer above the ground where it can be read. The changes in weight, with due allowance for rain (which is measured separately) and for any drainage (which is stored and measured below the lysimeter) is the amount lost by evaporation. A large mechanical balance, perhaps supporting several tons of container, is expensive, and so too is the structure in the ground that houses them. Weighing lysimeters are not, therefore, for general use, but they are one of the best methods of measuring evaporation.

Observing strict terminology, it is more correct to call lysimeters that are sealed and have no drainage *evapotranspirometers* and only to call those that drain lysimeters.

Measuring transpiration and evaporation from plants   It is sometimes the case that the largest contributor to water loss is transpiration. This can be measured directly by cutting through the stem of a plant, or the trunk of a tree, and immersing it in a tank of water. By measuring the fall of water level in the tank, plant uptake is sensed directly, any form of water level sensing technique being suitable (Chapter 10). But it is necessary to keep a watch on the health of the plant or tree and to be certain that it remains representative of the surrounding vegetation, particularly as regards *stomatal resistance* since the degree of openness of the leaves' stomata controls the rate of transpiration. This is difficult to achieve, however, because the plant in the tank has an unrestricted water supply and there may well be differences in transpiration from plants growing naturally in the ground, as their water supply varies with time. The method can only be used for a short time, as in the tank the plant cannot grow naturally for long.

An alternative is to allow the plant to remain in the ground and to measure the sap flow, a tracer of heat or chemical being injected into the trunk. In one such design, intended for larger trees, two thermocouple needle-probes are inserted into the sapwood, one above the other a few centimetres apart, the upper one being heated a few degrees. The difference in temperature between the two probes, resulting from heat being carried away by sap flow, gives an indication of the volumetric flow rate. As flow rate can vary round the circumference of the tree, some designs have up to eight pairs of differential temperature sensors (Steinberg *et al.* 1989, Smith & Allen 1996). Designs are also made for stems as small as 2–5 millimetres.

Measurements can also be made of the amount of rainwater lost by evaporation directly from the plant's surface. This is most usually done only for large vegetation such as forests; the rainfall above a forest canopy and the amount that reaches the forest floor (the *throughfall*), are measured, the difference between the two measurements being the amount lost from the trees by surface evaporation – the *interception loss*. Account must also be taken of that part of the rain that finds its way to the ground down the tree trunks – the *stem flow*.

The above-canopy rainfall is measured by installing a raingauge funnel on a mast or tower, level with the top of the forest, a tube feeding the water to a tipping-bucket raingauge at ground level (Chapter 8). The throughfall, arriving at ground level, is measured in one of two ways, depending on the type of forest. In dense plantations, large collectors of plastic sheeting are suspended below the trees on frames and are also attached to the tree trunks; these collectors direct the collected throughfall and stemflow to a large tipping bucket (Chapter 8). For sparse and natural forests, it is possible to use a cheaper method in which simple funnels and bottles are positioned on a random grid that is changed from time to time to avoid any bias. Stem flow can be measured in a similar simple way.

These techniques illustrate well the difficulty of measuring evaporation and why extremely roundabout and expensive ways may have to be used.

Measuring the flux of water vapour

While the previous methods estimate evaporation by measuring indirectly the net loss of liquid water to the atmosphere, direct measurement of the actual flux of water vapour requires the sensing of the rate of flow of the vapour from the ground into the atmosphere. There are several ways of doing this, but the two most used and most precise are the energy balance and the eddy correlation methods.

The Bowen-ratio (energy-balance) method   Water vapour and heat diffuse from where they are more concentrated, near to the evaporating surface, to where they are less concentrated, in the atmosphere above. The Bowen-ratio technique estimates the ratio between the *sensible heat flux* and the *latent heat flux*, the *Bowen ratio*, by taking measurements of temperature and humidity at different levels, thus giving their vertical gradients. Over forests, large towers are needed and complex instruments have to be moved up and down (to avoid instrument bias) to measure the small temperature and humidity gradients within and above forests. Simpler systems can be used over short vegetation because the gradients are larger since the flow is less turbulent and so there is less mixing. In the system illustrated (Fig. 6.2), two tubes, which can be separated vertically by a height of between 0.5 and 3 metres, draw air samples into a dew-point sensor (housed in the logger box at the base of the tripod), the samples being switched alternately every few minutes from one intake to the

Figure 6.2. The Bowen-ratio apparatus for measuring evaporation. On the end of each of the two horizontal arms is a thermocouple (a close-up is shown in Fig. 6.3), which senses the temperature gradient between the two heights. Beneath each arm are two air intakes (the button-like object, top arm); a fan draws air alternately from each level to a dew-point sensor in the logger box at the base of the mast, giving a measure of the humidity gradient. Out of view is a net radiometer and, below ground, a soil-heat-flux plate. An electronics unit in the box combines these six measurements to give an estimate of evaporation.

other, so giving measurements of the dew point, and thus the humidity, at the two heights; from these the gradient is calculated. The temperature at each height is also measured, by thermocouples (Fig. 6.3; Chapter 3). The readings are typically averaged over half to one hour while, over the same period, measurements are also made of net radiation and soil heat flux, their difference giving the energy available to power the sensible and latent heat fluxes. The Bowen ratio then allows this energy to be partitioned between the two fluxes to give a measure of evaporation (Lloyd *et al.* 1997).

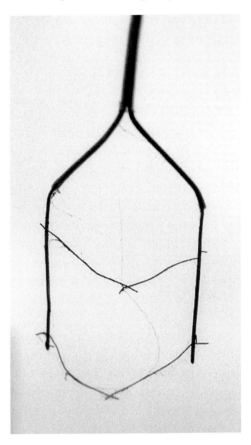

Figure 6.3. The thermocouple of a Bowen-ratio instrument. It is sufficiently small to be unaffected by radiative heating and so does not need a screen, allowing unimpeded measurements of temperature to be made with a rapid response time. Two (single) junctions are operated in parallel, in case one fails.

The eddy correlation method   Irregularities at the earth's surface interact with the wind to produce eddies which carry away water vapour and heat through the process of turbulent diffusion. Although at each height above the ground the mean wind direction is parallel to the surface, within turbulent eddies the movement is in three dimensions, changing rapidly with time and carrying water vapour and heat away from, or towards, the ground. This produces rapid fluctuations in the water vapour concentration in these eddies. Through this process, an upward flux of evaporated water occurs, since, generally, an upward movement of air correlates with a higher than average humidity – and conversely. The same logic applies to heat. It is, therefore, possible to measure the actual evaporation by taking simultaneous, rapid, instantaneous readings of the vertical component of the wind, and of humidity, the total evaporative loss being found by integrating all the rapid readings over a period of at least ten minutes and comparing this integrated value with a longer-term humidity average. By sensing the rapid temperature changes, as well as those in humidity, the sensible heat flux can also be measured directly in the same way, giving extra information against which the evaporative flux can be checked.

In practice these measurements are difficult to make, owing to the high speed at which the readings have to be taken and the small values of the changes involved. The reason that readings must be taken quickly is to include all the higher frequencies in the turbulent motion while still being able to sense much slower changes without undue drift in sensor response (Moore *et al.* 1976). The size and frequency of the eddies depend on the nature of the surface, the height of the measurement and the wind speed, but sensors must be able to take at least 10 readings a second, while also being able to see changes occurring over tens of minutes. It is also necessary for the instrument to be able to do the calculations in real time at least as fast as the readings are taken, although with modern electronics this is not difficult; 25 years ago it was less simple.

Suitable sensors to meet these requirements have already been mentioned in earlier chapters. The vertical fluctuations of wind speed are generally measured using a sonic anemometer (Fig. 6.4) although the use of pressure and thermal wind speed sensors is also a possibility, as are three-dimensional propeller anemometers (Chapter 5). All have been used at one time or another in eddy correlation and each has its advocates, although the sonic method is probably the best today. A cup anemometer, on top of the mast (Fig. 6.4), is used to measure the average horizontal wind speed.

Infrared radiation passing through air is absorbed by the water vapour it contains so, by detecting the amount of infrared energy absorbed, a measure of humidity is obtained. The response of such a sensor is instantaneous and the electronics is fast enough to sense the rapid changes of humidity in the turbulent air. The radiation bands used are those in the mid-infrared, from 1 to 6 microns. An infrared source, a light chopper and a detector constitute the sensor. The purpose of the chopper is produce an AC signal rather than a DC one, since drift is then less of a problem and the signal-to-noise ratio is also improved, allowing background noise to be filtered out. If absolute humidity was the requirement, the instrument would need to compensate for drift in the strength of the light source, in the sensitivity of the detector and in the electronics and for dust in the air, making it a more complex and expensive instrument. But what is required are the changes of humidity in the relatively short term, and techniques such as comparing the rapidly fluctuating readings with their running average achieve this with greater simplicity. It is also

Figure 6.4. The basic sensors of the eddy correlation method of measuring evaporation are the two vertical sets of tubes, seen at the left of the photograph, which measure the rapid fluctuations in vertical wind speed and humidity; the small horizontal tube measures fluctuations in temperature (it is fine enough not to be heated by radiation). In addition, the cup anemometer measures the long-term horizontal wind speed while, in the screen, sensors measure the average temperature and humidity. On the right, a net radiometer senses the available energy and, below ground, a soil-heat-flux plate measures the third energy flux, which together with the other measurements gives a direct measurement of evaporation.

possible to operate a capacitive, or wet-and-dry, humidity sensor, deployed in the miniature screen (Fig. 6.4), to give an accurate measure of humidity over the longer term, against which the readings from the infrared instrument can be referenced. A recent design puts the hygrometer at the centre of a three-dimensional ultrasonic wind sensor (Fig. 5.10), which measures exactly the same air that is sensed by the IR humidity sensor, rather than being slightly displaced to one side as in the current models.

The rapid changes in temperature are more easily and cheaply measured than those in humidity; thermocouples, small PRTs or thermistors are the most effective devices for this. All three can follow rapid fluctuations in temperature, fulfilling the same rôle for sensible-heat-flux measurement as the humidity sensor does for latent heat flux. A slower-responding temperature sensor, such as a larger PRT, is operated in the miniature screen to give a long-term reference.

As a check on the performance of the whole instrument and its calculations, a net radiometer and a soil-heat-flux plate can be operated close by, allowing the overall energy balance to be checked – the available energy measured by the net radiometer should be the same as the total of the sensible heat flux and latent heat flux, measured by the eddy correlation instrument, and the energy measured by the soil-heat-flux plate.

The accuracy of such an instrument is from $\pm 15\%$ to $\pm 20\%$ of the actual evaporation for hourly values and over longer periods $\pm 10\%$.

The method is also applicable to the measurement of other fluxes, in particular the carbon dioxide flux. This is of considerable current interest because through the measurement of the CO_2 flux over different vegetation and surfaces, such as the sea, it is possible to determine whether the plants or water are a net sink of the gas, or an emitter (Gash & Nobre 1997, Grace *et al.* 1995, Moncrieff *et al.* 1997). It is, however, more difficult to measure the CO_2 flux than that of water vapour because the concentration of the gas is much less (about 1%). Since, like water vapour, CO_2 also absorbs infrared radiation, its measurement is simply a matter of selecting a different frequency band to detect the CO_2. Because of the lower concentration of the gas, however, a more complex sensor system has to be used and this is only currently available as a rather bulky laboratory instrument, unsuitable for mounting in the open close to a wind sensor. To overcome this problem, air is drawn down a tube, from close to the sonic anemometer, into the sensor unit housed a few metres away (in a similar way to the air in the Bowen-ratio system, Fig. 6.2). But as it takes time for the air to reach the sensor, a correction has to be made to ensure that the wind and CO_2 measurements are in synchronism. Although it might be thought that fluctuations in the CO_2 level would be lost as the air flowed along several metres of tubing, this is not the case for the lower frequencies, although the higher frequencies are smoothed out through turbulent mixing in the tube. This is of no serious consequence, however, since it is at the lower frequencies that most of the transfer occurs, and in addition some correction can be made for the lost higher frequencies. There is also a slight advantage in the smoothing effect of the tube in that the small temperature fluctuations in the air are also smoothed out, this being useful since temperature fluctuations, through their effect on air density, affect both CO_2 and water vapour fluxes. The same method can equally well be used to measure the fluctuations of water vapour, and some eddy correlation instruments use this *closed-path* construction instead of the more traditional *open sensor* illustrated in Fig. 6.4.

However, the high cost of these systems and the need for experienced operators and for expert interpretation of the results restrict the use of the eddy correlation method to specialised applications (Gash *et al.* 1991, Gash & Nobre 1997). It must be of some concern that measurements from instruments of this type, in inexperienced hands, could be misinterpreted.

Estimating evaporation

Because of the cost and complexity of many of the foregoing methods, ways have been sought of estimating evaporation by making measurements of other variables and deducing the evaporation from them. From the strict scientific viewpoint these methods may not be ideal, but they can be much more economical than the precise methods discussed above, and they have a wide and useful place in the everyday world.

To be clear exactly what such techniques are estimating, several *standard rates of evaporation* have been defined. Penman (1948) developed the idea of *potential evaporation* – the amount of water evaporated (per unit area, per unit time) from an extensive area of a free water surface. Although it takes no account of energy exchanges within the water or of turbulent transfer above, the concept bears some relationship to actuality. The idea of *potential evapotranspiration* – the maximum amount of water that can be lost, as water vapour, from a large area of vegetation when the soil is saturated (Gangopadhyaya *et al.* 1966) – was a later development. It was based on experimental evidence (at the time of the development of the concept) which suggested that the type of vegetation had little effect on evaporation, so long as evaporation was limited by the energy available at the surface and not by the availability of water. However, it became clear that the vegetation did in fact have an effect and the concept of *reference crop evapotranspiration* was introduced. In Penman's original work, the evaporation was from turf. There are now a number of variations, but the method based on energy balance, of which Penman's is the classical example, is the only one considered here.

The Penman method   Penman's method of estimating potential evapotranspiration requires the measurement of energy input, air temperature, humidity and wind speed. In one version of the calculation the energy input information is obtained from the direct measurement of net radiation; in another, the net radiation is estimated from solar radiation or sunshine hours. The direct measurement of net radiation gives much better results since the measurement is more accurate than those derived indirectly from sunshine records.

The Penman equation takes the form

$$E = \left(\frac{\Delta}{\Delta + \gamma}\right) H_T + \left(\frac{\gamma}{\Delta + \gamma}\right) E_A$$

where H_T is the *heat budget* and E_A is the *aerodynamic term*. Δ and γ are *efficiency factors* controlling the relative effectiveness of the energy supply and ventilation (Penman 1948); Δ is the slope of the temperature versus saturation vapour pressure curve and γ is the *psychrometric constant* (Chapter 3). The evaporation is thus calculated in two parts, the heat budget and the aerodynamic term, both expressed in equivalent millimetres of water evaporated in a given time.

The heat budget is calculated as R_N/L, that is the measured or estimated net energy R_N (with an assumed proportion taken up by the sensible and soil heat flux) received over the period of the calculation, usually an hour or a day – depending on the interval for which the net data are available – divided by L, the latent heat of vaporisation of water. Rijtema (1965) showed that providing the net radiation is measured over the vegetation concerned, the method provides an estimate of *reference crop evaporation* rather than potential evaporation.

The aerodynamic term is expressed as $E_A = 0.35 \, (1 + u/100) \, (e_a - e_d) \, P_c$, where u is the wind run in metres per day at a height of two metres and e_a and e_d are the saturation vapour pressure and the actual vapour pressure, to which a correction, P_c, has to be applied for altitude: $P_c = 1 + h/20\,000$, h being the altitude in metres (McCulloch 1965).

The Penman equation as shown above is the simplest physically based model which can be used to estimate evaporation. Over the years, other workers have refined the method, notably Monteith (1965), while Thom & Oliver (1977) modified the original formula to include extra terms with a physical basis which could be changed empirically, to give an estimate of evaporation from different crops and for use over different vegetation including forests.

The instrumentation required for the estimation of evaporation by the Penman method, which needs the measurement of net radiation, temperature, humidity and wind speed, has already been described in earlier chapters. However, one of the motivations for developing the first automatic weather stations (AWSs) in the 1960s and 70s (Strangeways 1972, Strangeways & Smith

Figure 6.5. To estimate evaporation by the Penman method, an AWS has sensors for measuring net radiation (left lower arm), temperature and humidity (in the screen, right lower arm) and wind speed (right upper arm). To provide additional climatological background, there are also sensors for solar radiation (top of the mast) and wind direction (left upper arm) and, out of view, a raingauge. Soil temperatures may also be measured, while for meteorological use there will be a barometric sensor. The antenna is for telemetering the measurements via Meteosat (Chapter 12).

1985) was to measure this complement of Penman variables. Rainfall was also included to give a measure of the precipitation input. In addition solar radiation, wind direction and barometric pressure may be included to form a complete *climatological station* (Fig. 6.5) and to these can be added soil temperature, soil-heat-flux plates and any other sensors that may be needed for specialised applications. AWSs are multipurpose instruments that are increasingly being used instead of what is commonly called the *met. enclosure*, a fenced-off area that houses a collection of old-style instruments; see Fig. 6.6 (Strangeways 1995). It was data from such conventional enclosures, however, that Penman used when developing his methods in the 1940s and 50s.

Other methods of estimating evaporation, which make more empirical assumptions than the Penman method, have been developed; these rely on the measurement of fewer variables and thus on simpler instrumentation.

Figure 6.6. A 'met. enclosure'. In the 1950s, before AWSs were generally available, Penman developed his technique of estimating evaporation using conventional met. enclosure data. Such sites are still the mainstay of most of the NWSs for synoptic data collection (i.e. for weather forecasting) and they are also still widely used for many research projects. The site shown here is in Java, and in addition to a wide selection of raingauge types and an AWS, a conventional Stevenson screen and anemometer (right) can be seen. Out of shot is a selection of radiation-measuring instruments and a wind-direction vane.

Radiation methods   Penman (1956) commented that, in Europe at least, the heat-budget part of his equation was usually at least four to five times greater than the aerodynamic part. A fair approximation of evaporation might thus be obtained by omitting the second term but making some notional allowance for it (Priestley & Taylor 1972). This approach means that it is then not necessary to measure humidity, just radiation. However, Shuttleworth (1979) and Shuttleworth & Calder (1979) showed that the method did not work correctly for forests, although by changing one of the factors in the equation (we will not pursue this here) the method might give an estimation of transpiration. It is only a second best to the original Penman method, but of the various simplifications probably the best.

Humidity methods   Despite Penman's observation that the radiation term is by far the larger of the two, it is nevertheless possible to relate just the second term to evaporation. This involves the vapour pressure deficit $e_a - e_d$ and the wind speed: the dryer the air and the windier the day the greater the evaporation. The method was advocated by John Dalton (1802), who also did experiments with lysimeters (see earlier) and with evaporation pans (see later). But where such humidity and wind measurements exist, it is better to use the full Penman method, using the estimated net radiation (or the measured net radiation if possible). Humidity methods, therefore, are not particularly practical or recommended.

Temperature methods   There are many empirical formulae that try to relate evaporation to temperature alone, all based on the assumption that the two terms in the Penman equation probably have some loose correlation with temperature. Since the first term of the equation is usually by far the greater, it is the relation between temperature and radiation that is important; but there is a lag between the annual temperature cycle relative to radiation and this is a problem unless allowance is made for it. The Blaney–Criddle equation (1950) is one of the better known of this type and was designed to give daily estimates of evaporation averaged over a month. It is expressed as

$$E = c_u d_1 (0.46T + 8),$$

where c_u is a *consumptive factor*, best determined for the particular site con-cerned, d_1 is the fraction of daylight hours occurring in the month and T is the average temperature (in °C; the constants reflect the fact that the originals were derived using °F, as is often the case in equations of this type). However, the easier availability of net radiation data today, from AWSs, means that this approach is not often used.

These alternative, simpler methods can of course help predict roughly what the evaporation might be in the future using available past data from old-style instruments; it is often the case that engineers have to make decisions, and design structures, with nothing more than crude past data to work on. Such estimation methods may then have a place, but not when instruments are to be set up to measure evaporation in the future. They are worth discussing, however, as they put the Penman method in its rightful place at the top of the estimation methods.

Evaporation pans and atmometers

While strictly speaking these devices are further methods of estimating evapor-ation, they will be treated here in more detail because of their extensive use. Evaporation pans in particular are undoubtedly the most common means in use today of estimating or measuring evaporation and so need special con-sideration. Like raingauges, but to a lesser extent, evaporation pans have a history going back at least to the eighteenth century. In Southport, near to Liverpool, Dobson (1777) operated a small pan of 12 inches (30 cm) diameter by 6 inches deep, with a raingauge of the same size nearby. Based on a recent estimate using the Penman method, it seems that his measurements of evapor-ation were too high. John Dalton later (probably around 1785) used Dobson's method at Kendal in the Lake District, for 82 days (May to June inclusive) and measured 5.41 inches (13.7 cm) of evaporation.

Atmometers   Atmometers are not widely used, but they need to be mentioned, their attraction being low cost and simplicity. They operate by measuring the loss of water from a wetted surface (Gangopadhyaya *et al.* 1966). Although there are several designs, the Piche instrument is probably the most common, consisting of a graduated tube, closed at the top and with a filter paper disc fixed to the bottom. As the water evaporates from the paper, the level falls in the tube and is read periodically. In one design the level is recorded on a chart. The energy for evaporation is derived from the radiation falling on the disc and from the heat conducted through the water supply from the exposed surfaces of the container. So how atmometers are sited greatly affects how they perform, and there is no agreed satisfactory way of exposing them. They can also be affected by dust on the evaporating surfaces and by the surroundings. Although a well-maintained instrument can sometimes have a passable correlation with potential evaporation, a different relationship is to be expected for each situation and any idea of universality is an illusion.

What might be described as a miniature evaporation pan, although in concept it is an atmometer, is the *recording evaporation balance*. It consists of an open water container with an area of $250 \, cm^2$ (17.84 cm diameter) and a depth of about 3 cm; this is placed on a mechanical spring balance, which records the change in weight on a paper chart as the water evaporates. It is not a very common instrument and will suffer from the same problems as other atmometers. No doubt it has its advocates and used intelligently may give some rough indication of evaporation.

Evaporation pans   Probably the oldest, and still the most widely used, method of estimating evaporation is the direct one of relating potential evaporation to the evaporation from tanks of water exposed to the elements. Gangapadhyaya (1966) described 27 different designs of pan and suggested that the list was probably far from complete. In its multiplicity, it is probably only exceeded by raingauge diversity. This alone is a drawback since it prevents meaningful intercomparisons between sites, but the main problem is whether pans give a measurement that can be related to potential evaporation, or indeed to any type of evaporation.

The energy exchange of pans will differ from that of the ground and the vegetation around them. In deep pans, such as the British Meteorological Office's standard tank (Fig. 6.7), which is 1.83 m square by 610 mm deep and is installed with the top 75 mm above the ground, the stored energy will tend to be greater than that of the vegetation so that the surface temperature will be lower in the day, when most evaporation takes place, and higher at night. Russia has two designs of pan, one being 618 mm in diameter by 600 mm deep,

Figure 6.7. UK evaporation pans are sunk into the ground. Changes in water level are measured manually, using the hook gauge in the left-hand corner of the pan.

set in the ground, with the top 75 mm above ground; the second is very large, with a diameter of 5 m and a depth of 2 m.

If a pan is raised above the ground and is shallow, as is the Class A pan (Fig. 6.8), there will be radiation exchange through the exposed sides, and this will cause a response different from that of the ground and plants. Performance will also be affected by the colour of the pan, a darker surface absorbing more solar radiation, and the colour of the water (for example if algae are allowed to grow in it). But pans exposed above the ground are less expensive and are easier to instal, less debris will blow into them and less rain will splash into them, while leaks are more easily detected and corrected.

As with all instruments, exposure is an important factor, the conditions needed for an evaporation pan being similar to those for raingauges. A raingauge must also be operated nearby, to allow corrections to be made for rainfall falling into the pan. Birds and larger animals might drink from pans, and must be stopped by fences or gratings over the top. Any fence around the pan should be open so as not to obstruct the wind. If evaporation from a lake is to be measured, readings from a pan floating in the lake will more closely approximate to the lake's own evaporation than measurements from a pan on land, but operational difficulties are considerable.

Like many of the older instruments that are still widely used, much work has been done in comparing the relative performance of the different pans and, as a result, the Class A pan was selected by WMO as the world standard.

Figure 6.8. The Class A evaporation pan, chosen by the WMO as the world standard, is smaller than the UK model and sits above the ground on a frame. In this photograph, taken in the state of Rajasthan in India during the monsoon, the hook gauge for measuring water depth is being read. Despite the doubtfulness of evaporation-pan measurements, this is the most common method, worldwide, of getting an estimate of evaporation.

The principle of operating a pan is similar for all types. Usually once a day the level of the water in the pan is measured with a *hook gauge* in a small stilling well in the tank (Figs. 6.7, 6.8). The hook (a pointed rod) is attached to a graduated scale with a vernier allowing reading to 0.1 mm, the hook and scale being turned up and down by a rack and pinion movement. It is moved up until the (upturned) point just cuts the surface of the water (this is easier to detect than lowering the point until it meets the water). The change in reading since the previous occasion, with due allowance for precipitation, gives the amount of evaporation that has occurred from the pan. The water in the pan is kept between 5 and 7.5 cm from the top of the tank, either by adding or removing water from time to time. This is done immediately after taking a reading, followed by a second reading to establish the new depth. An alternative is to use a fixed level, the amount of water that has to be added or removed being measured in a graduated cylinder (with one-hundredth the area of the pan).

Extra measurements may be made to help interpret the pan measurements more meaningfully, including wind speed at pan level and the maximum and minimum temperature of the water – measured either by thermometers floating on the water or lying on the bottom. A further enhancement is the addition of an automatic water level recorder in place of the manual hook gauge. These are usually external units in which a float turns a potentiometer (Chapter 10) through a rack and pinion mechanism, giving an analogue output for logging.

References

Biswas, A. K. (1970) *History of Hydrology.* North-Holland, Amsterdam and London.

Blaney, H. F. & Criddle, E. T. (1950) Determining water requirements in irrigated areas from climatological and irrigation data. USDA (SCS) TP-96, 48p.

Dalton, J. (1802) Experiments and observations to determine whether the quantity of rain and dew is equal to the quantity of water carried off by the rivers and raised by evaporation; with an enquiry into the origin of springs. *Memoirs, Literary and Philosophical Society of Manchester,* **5**, part 2, 346–72.

Dobson, D. (1777) Observations on the annual evaporation at Liverpool in Lancashire; and on evaporation considered as a test of the dryness of the atmosphere. *Phil. Trans. Roy. Soc. London,* **67**, 244–59.

Gangopadhyaya, M., Hurbeck, G. E., Nordenson, T. J., Omar, M. H. & Uryvayev, V. A. (1966) Measurement and estimation of evaporation and evapotranspiration. WMO, Geneva, Technical note No. 83.

Gash, J. H. C., Wallace, J. S., Lloyd, C. R. & Dolman, A. J. (1991) Measurements of evaporation from fallow Sahelian savannah at the start of the dry season. *Quart. J. Roy. Met. Soc.,* **117**, 749–60.

Gash, J. H. C. & Nobre, C. A. (1997) Climatic effects of Amazonian deforestation: Some results from ABRACOS. *Bull. Am. Met. Soc.* 1997, 823–30. (ABRACOS is the acronym for the Anglo-Brazilian Amazonian Climate Observational Study.)

Grace, J. *et al.* (1995) Carbon dioxide uptake by an undisturbed tropical rain forest in southwest Amazonia, *1992–1996. Science,* **270**, 778–80.

Law, F. (1956) The effect of afforestation upon the yield of water catchment areas. *J. Brit. Waterworks Ass.,* **38**, 489–94.

Lloyd, C. R. *et al.* (1997) An intercomparison of surface flux measurements during HAPEX-Sahel. *J. Hydrol.,* 188–9, 385–99.

McCulloch, J. S. G. (1965) Tables for the rapid computation of the Penman estimate of evaporation. *J. E. African Agric. Forestry,* **84**, 286.

Moncrieff, J. B. *et al.* (1997) A system to measure fluxes of momentum, sensible heat, water vapour and carbon dioxide. *J. Hydrol.* (HAPEX-Sahel special issue) **188–9**, 589–611.

Monteith, J. L. (1965) Evaporation and the environment. *Symp. Soc. Expl. Biol.,* **19**, 205p.

Moore, C. J., McNeil, D. D. & Shuttleworth, W. J. (1976) A review of existing eddy-correlation sensors. Institute of Hydrology Report No. 32.

Penman, H. L. (1948) Natural evaporation from open water, bare soil and grass. *Proc. Roy. Soc. London,* **A193**, 120p.

Penman, H. L. (1956) Evaporation: an introductory survey. *Netherlands J. Agr. Sci.* **1**, 9–29.

Priestley, C. H. B. & Taylor, R. J. (1972) On the assessment of surface heat flux and evaporation using large scale parameters. *Mon. Weather Rev.,* **100**, 81.

Ritjema, P. E. (1965) An analysis of actual evaporation. Agricultural Research Report 659, Pudoc, Wageningen, 107p.

Rodda, J. C., Downing, R. A. & Law, F. M. (1976) *Systematic Hydrology.* Butterworths, London.

Shuttleworth, W. J. (1979) Evaporation. Institute of Hydrology Report 56.

Shuttleworth, W. J. & Calder, I. R. (1979) Has the Priestley–Taylor equation any relevance to forest evaporation? *J. Appl. Met.,* **18**, 639–46.

Smith, M. D. & Allen, S. J. (1996) Measurement of sap flow in plant stems. *J. Expt. Bot.*, **47**, 1833–44.

Steinberg, S. L., van Bavel, C. H. M. & McFarland, M. J. (1989) A gauge to measure mass flow in stems and trunks of woody plants. *J. Am. Soc. Hort. Sci.*, **11** (4), 466–72.

Strangeways, I. C. (1972) Automatic weather stations for network operation. *Weather*, **27**, 403–8.

Strangeways, I. C. & Smith, S. W. (1985) Development and use of automatic weather stations. *Weather*, **40**, 277–85.

Strangeways, I. C. (1995) Back to basics: the met. enclosure: Part 1 – its background. *Weather*, **50**, 182–8.

Tom, A. S. & Oliver, H. R. (1977) On Penman's equation for estimating regional evaporation. *Quart. J. Roy. Met. Soc.*, **103**, 345p.

7

Barometric pressure

The variable

History

Mining engineers in Italy in the seventeenth century, finding it impossible to draw water by single-stage suction-pump to a height of more than about 10 metres, sought an explanation; this also tied in with the question posed by Aristotle as to whether a vacuum could exist in nature. In 1643, Torricelli performed his classic experiment with mercury in an inverted tube and showed that the reason for the pumping problem was that a vacuum formed over the water if suction continued to be applied beyond the height of 10 metres. The atmosphere has weight, and it is the pressure of this weight that supports the liquid column to a height such that its pressure equals atmospheric pressure. The word *barometer* was coined from the Greek root *baros*, meaning weight. Five years later Pascal and Perrier showed that the pressure was less at the top of a mountain than at its base, and it was soon realised that barometric pressure also varied with the weather.

At the same time, Robert Boyle was studying in Italy, reading Galileo's writings (which included references to air pressure), and he became aware of the Torricelli experiment. When he returned to England he brought the idea of the barometer with him, and went on to formulate his theories about the relationship between the pressure, temperature and volume of gases – Boyle's laws.

The reason why barometric pressure varies with time and place is that various processes change the amount of air molecules in the atmosphere immediately overhead, increasing or reducing its weight. Cooling or warming of the air is the main cause of this, and it may happen by differing amounts at different heights up through the atmospheric column. It could also be that at some height the air cools while at others it warms. It is the integrated effect over the full height that controls the total weight. Thus a barometer at one height may indicate a rise while at another height, in the same column, a fall. The general rule is that pressure falls in response to rising temperatures aloft, and rises in response to cooling (Burton 1993). Changes in temperature can occur in a static column of air, for example owing to solar heating near to the ground. But temperature changes also occur when a cold or warm front passes. Changes in pressure are also influenced by the more complex processes of *convergence* and *divergence*, in which air molecules are added to or extracted from the column of air overhead, usually at altitudes of from 8000 to 12 000 metres (Young 1994).

Units

The SI unit of pressure is the newton per square metre ($N\,m^{-2}$), to which the name Pascal has been given (Pa). For meteorology, however, the millibar is used (one-thousandth of a bar) and, although the international symbol for this is mbar, the meteorological community uses mb. The relationship between the pascal and millibars is that $1\,mb = 100$ Pa. So $1000\,mb = 10^5$ Pa, usually expressed as 1000 hectopascals (hPa), making hectopascals interchangeable with millibars. It is probable that hectopascals will be adopted as the unit for measuring atmospheric pressure in the future and this is already happening.

The newton is the unit of force. As $1\,newton = 0.102\,kg$, $1\,bar = $ a pressure of $10\,200\,kg\,m^{-2}$ ($14.5038\,lb\,in^{-2}$). But 1 standard atmosphere (atm) $= 1.0132$ bar ($1013.2\,mb$ or $14.7\,lb\,in^{-2}$). Therefore

$$1\ bar = 0.9869\ atm.$$

Translated into the equivalent height of a mercury column at mean sea level we have

$$1\ bar = 750.06\ mm\ (29.53\ in),$$

and

$$1\ atm = 760.00\ mm\ (29.92\ in),$$

both at $0\,°C$. The equivalent of 1 atm for a water column is $10.33\,m$ ($33.9\,ft$) at $+4\,°C$.)

Mercury barometers

From the second half of the seventeenth century, mercury barometers became commonplace, as much for domestic use and decoration as for science. They are now all collectable antiques. Initially they were used more as altimeters, but their meteorological significance soon became recognised and studied.

The mercury barometer is still amongst the most accurate means of measuring barometric pressure and is the same in principle as that first made by Torricelli 354 years ago, although refinements have been added in the intervening centuries. Nevertheless, today the aneroid barometer (see below) can now equal the mercury barometer and has largely replaced it, although carefully maintained mercury instruments are still used as primary standards.

Barometers have been made in great numbers and in considerable artistic variety (Banfield 1976). In one design, developed before aneroid barometers were invented, the movement of the mercury in the cistern is measured using a float and counterbalance that turn a wheel of 8 to 12 inches diameter, to which is attached a pointer that moves round a dial. This design is commonly ascribed to Robert Hooke who demonstrated it in 1664. It later became known as the *banjo* barometer and was one of the cheapest and most popular of the mercury barometers made over the years (McConnell 1994).

The Kew-pattern barometer

Two present-day designs exemplify the mercury barometer. The *Kew-pattern* barometer, used throughout the world for synoptic and climatological applications, differs from the original seventeenth century design in small but important ways. The cistern is sealed against dust by a leather washer, with small holes to allow air pressure changes to diffuse through (Fig. 7.1). The bore of the cistern is precisely 5 cm while the glass tube emerging from the cistern has a bore of 1.6 mm, widening to provide an air trap, then narrowing again before finally opening up at the top to 8 mm, giving it a broad flat top for better reading. The mercury is triple distilled and filtered to exclude impurities that would affect its density and also to ensure the formation of a clean meniscus that does not stick to the sides of the tube.

At the meniscus level, the surrounding brass tube is engraved in millimetres or inches of mercury, or in millibars, with a sliding vernier scale to allow readings to 0.1 mb. The advantage of this design is that only one setting needs to be made – that of the vernier – because the bores of the cistern and tube are precisely known, allowing the vernier to be designed to take account of the change of mercury level in the cistern and so making it unnecessary to adjust the mercury level in the cistern to a fixed reference point.

There are two models, one for normal use from sea level up to 450 m (1500 ft), covering the pressure range 870 to 1060 hPa, and a second model for use up to 1070 m (3500 ft) covering the range 780 to 1060 hPa.

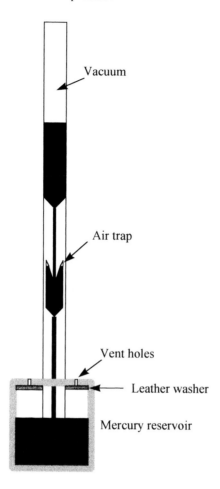

Figure 7.1. The Kew-pattern barometer is shown here schematically, and not to scale (the length of the tube being foreshortened), so as to show the small but important parts – the narrowing of the main tube to minimise the effects of rapid pressure changes, the air trap, the vent holes and a leather washer to keep the mercury dust-free.

The Fortin barometer

Another commonly used mercury barometer of today remains unchanged in concept since it was designed by Nicholas Fortin in 1810. It differs from the Kew model in requiring two adjustments when taking a reading. The difference lies in the fact that the mercury level in the cistern is visible through a glass tube in which is fixed a downward-pointing needle (Fig. 7.2(a),(b)). The base of the cistern is made of flexible leather supported by a plate that can be screwed up and down, adjusting the level of the mercury so that it coincides exactly with the

(*b*)

(*a*)

Figure 7.2. (*a*) The mercury container of a Fortin barometer is of flexible leather (within the brass tube) and is adjusted by the lower knob until the mercury is level with the fiducial point. This compensates for the change in mercury level in the container caused by the changes in level in the main tube due to variations in barometric pressure. (*b*) The *fiducial point* of a Fortin barometer (right, within the glass tube), against which the level of mercury is adjusted to keep it constant.

point of the needle, at the so-called *fiducial point*. This gives a precise reference, against which the level of the mercury in the main tube is measured by a vernier (Fig. 7.3(*a*), (*b*)), just as with the Kew model except that the vernier scale does not have to be corrected to allow for the level change in the cistern. Like the Kew design, the Fortin barometer is made in two models, one for low altitudes and

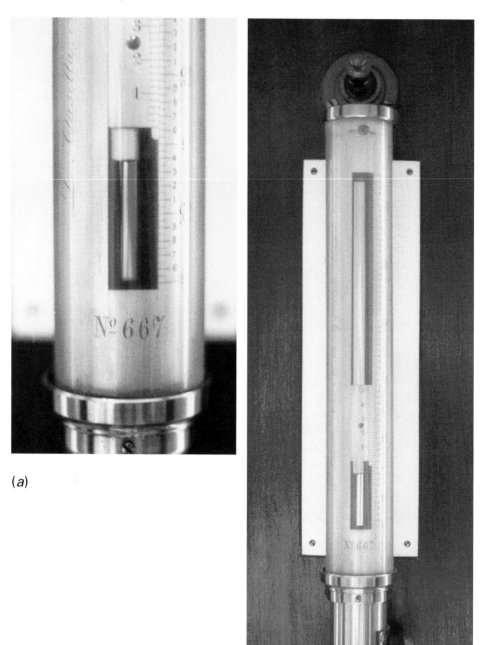

(a)

(b)

Figure 7.3. (*a*) At the top of a Fortin barometer, the level of the mercury meniscus is measured by a vernier scale, adjusted by the knob on the right; see (*b*).

one for higher altitudes. It has the advantage that it is somewhat more transportable than the Kew design, but nevertheless the transport of any of these instruments has never been easy. Indeed many early design modifications were attempts to overcome this problem (Chapter 13), but it was not until the invention of the aneroid barometer that the transport problem was finally solved.

Errors and corrections

Barometers are designed to give the correct reading at $0\,^{\circ}$C. As the density of mercury, and thus its weight, changes with temperature, it is necessary to apply a correction to measurements made at any other temperature. For this purpose, barometers are supplied with a thermometer fixed to the case, giving as close as possible an indication of the temperature of the mercury column. To avoid errors, it is important that barometers are not exposed to direct sunlight or other source of heat and that they change only slowly in temperature, thereby avoiding temperature gradients within them.

Similarly, if the acceleration due to gravity varies from the standard reference value, $980.665\,\mathrm{cm\,s^{-2}}$, at the point of observation, a second correction must be applied, since the value of g also affects the weight of the mercury.

If the vacuum above the mercury meniscus becomes contaminated, for example through the introduction of small amounts of air or other gas, readings will be in error to an unknown amount depending on the degree of contamination.

The vernier of both types reads to 0.1 mb. Accuracy will depend on the care taken over the above errors.

Aneroid barometers

A brief history

Lucien Vidie invented the aneroid ('without liquid') barometer in 1843; this started the decline of the rather expensive mercury instrument, since the aneroid barometer was not only cheaper but also compact and portable, if not quite so accurate initially. Although French, Vidie developed his instrument in London, tested it at the top of St Paul's cathedral and also on train journeys from London to Chester and from Preston to Carlisle. Vidie found it difficult to sell his new barometer in France but won a Council Medal when he exhibited it at the Great Exhibition of London in 1848. But, having patented his design in 1844, he then had legal squabbles with Bourdon, a Paris instrument maker, finally winning damages in 1861. Bourdon's gauge is of course well known to engineers; it is used for measuring the high press-ures of steam and water in industrial machines such as steam engines, and takes the form of a spiral tube that uncurls as the pressure increases inside it.

As with mercury barometers, many aneroid models have been made for domestic use since its invention, probably more so than mercury owing to their lower cost. The commonest design to be seen in houses looks very much like the banjo mercury barometer.

Principles

The basic aneroid sensing element is a circular, shallow, thin corrugated metal capsule about 5 cm in diameter with the air removed from it, variations in barometric pressure compressing or relaxing the box causing it to contract or expand. One side of the box is fixed, the other is movable, the pressure of the air being balanced by a spring that prevents it from collapsing under the pressure of the atmosphere. In its simplest form, the movement of the capsule is magnified by levers which move a pointer around a dial. Its range is typically from 900 to 1060 hPa in 1.0 or 0.5 hPa steps.

Temperature variations cause small dimensional changes in the capsule and in the linkage mechanisms. To compensate for this, two methods are used. In the simpler, a small amount of air is left in the capsule; the air expands as the temperature increases, this being arranged to act in opposition to the un-wanted temperature-induced movements. Alternatively, one of the levers in the magnification mechanism can be a bimetal strip, which bends in compensation as the temperature changes. With modern designs and materials and with precision engineering, aneroid instruments can now equal the accuracy of mercury barometers and so have largely replaced them, even in professional meteorology.

Barographs

An aneroid barograph is an aneroid barometer in which the expansion and contraction of the capsule is recorded by a pen on a spring-driven or electri-cally driven paper chart. This requires modifications to the non-recording model, notably the stacking in series of several aneroid capsules to give more movement, so requiring less mechanical magnification. Magnification reduces the power available at the pen, which must overcome friction against the paper. Too much magnification causes the pen to jump in steps instead of recording each small change of pressure (a 'lively' record is the aim). A typical

Figure 7.4. A barograph converts the expansion and contraction of a stack of aneroid capsules (eight in this case) into a movement of a pen, via a series of levers. The one illustrated here, along with the Fortin barometer in the previous figure, is at the Royal Meteorological Society in Reading.

barograph will use from 8 to 14 capsules (Fig. 7.4), while a simple non-recording barometer will use just one. The range covered is usually 950 to 1050 hPa. The accuracy is about ± 1.0 hPa, but its exact value depends on pen and lever friction.

Errors and corrections

Temperature changes are compensated for automatically in aneroid instruments and so corrections are not necessary at room temperature. Unlike the mercury barometer, the balancing of atmospheric pressure is against a spring, not against the weight of mercury (which is dependent on gravity) and so no correction is required for variations in g. In determining mean sea-level pressure, the only correction that need be applied is for altitude, and for this the height of the barometer should be known to within one metre. However, with the more precise instruments, individual calibrations are carried out, and a card indicating corrections, unique to that instrument, is supplied with each barometer. Leakage can occur in the capsules, but it is rare and can be detected through periodic calibration checks.

When an accuracy is quoted by a manufacturer, this will normally round up, in one combined figure, all the usual suspects – non-linearity, hysteresis and repeatability – as well as any deviations due to imperfect temperature-compensation. There is also the additional matter of long-term stability, although, like leakage, this can be detected by periodic checks.

Electronic barometers

Aneroid-capsule type

Manual measurement   Although now being replaced by electronic sensors, aneroid barometers combining manual and electrical methods to measure the movement of aneroid capsules have been used in the recent past. A stack of about three capsules is arranged to move an electrical contact, in line with a second contact fixed to a micrometer screw. The full range of atmospheric pressure change produces a contact movement of about 1 to 1.5 mm, its position being determined by turning the micrometer screw until the two contacts just touch; the turning of the micrometer is sensed electrically, producing a digital signal. But it needs an operator. The accuracy of such methods is around ± 0.3 hPa, although the discrimination is higher, at 0.02 hPa.

Capacitive measurement   The majority of automatic electronic pressure sensors now use newer methods, but some have been designed around the aneroid principle, the physical movement of the capsule being sensed by capacitive plates installed within the vacuum cavity. In one such design, three separate identical units are used, their outputs being averaged to give a more accurate mean reading. Errors arise largely from the aneroid capsule rather than from the capacitive sensing element, and so instruments of this type have an accuracy akin to other precision aneroid methods.

Flexing diaphragm type

Sensed by straingauges   The aneroid capsule is of course a flexing diaphragm, but with the arrival of microelectronic fabrication techniques came the ability to make silicon diaphragms that flex under pressure (Chapter 10). Onto these thin wafers, using microelectronic etching techniques, are formed straingauge bridges that give an analogue voltage output proportional to the deflection of the diaphragm, which is amplified and either displayed visually or automatically logged. The accuracy for most pressure sensors of this type is, at best, about $\pm 0.1\%$ of the range covered. Thus if the range is 950 to 1050 hPa, 0.1% is ± 0.1 hPa. But this is the best achievable, and a more realistic accuracy, and one typically quoted, is between ± 0.3 to 1.0 hPa – comparable with aneroid sensors.

Sensed capacitively   Instead of straingauges, the deflection of the silicon diaphragm is now frequently measured capacitively: a flexing diaphragm and a fixed plate within the reference vacuum cavity have a small capacitance, the value of which varies as the diaphragm flexes and the distance between the plates changes (see also river-level pressure sensors, Chapter 10). This capacitor forms part of an oscillator circuit, the frequency of which is related to the degree of flexing and thus to the pressure. Quartz is sometimes used as the diaphragm since it has the possible advantage over traditional diaphragm materials of better elasticity, thereby improving repeatability and hysteresis performance. Ceramic materials are also used, with little difference in the accuracy achieved, a value of ± 0.3 hPa again being typical although some claim that ± 0.1 hPa is possible. It would seem, however, that ± 0.3 hPa is about the practical limit to the accuracy of all but mercury barometers – and the vibrating cylinder sensor, which will now be discussed.

Vibrating-cylinder type

A quite different technique is used in the *vibrating-cylinder pressure sensor.* Originally developed for use in aircraft and rockets as altimeters and for sensing pressures associated with their engines and fuels, the method has been modified to measure barometric pressure. Figure 7.5 shows a cross-section of the sensor, the space between the two concentric cylinders being a near vacuum and the inside of the thin-walled inner cylinder of magnetic alloy being open to the atmosphere. Three drive coils half-way along the sensor cause the inner cylinder to vibrate in a hoop mode (that is, not longitudinally). Changes in pressure change the hoop stresses and thus the natural frequency of vibration of the cylinder; these changes are detected by electromagnetic pick-up coils. Their signal is also fed back to the drive coils to maintain the resonance of the cylinder. But the frequency is also influenced by the density of the air, although correction for this is easily made by measuring the temperature of the sensor. The sensors are calibrated by measuring the frequency of the cylinder at up to 10 different pressures and temperatures, producing a unique calibration table for each sensor.

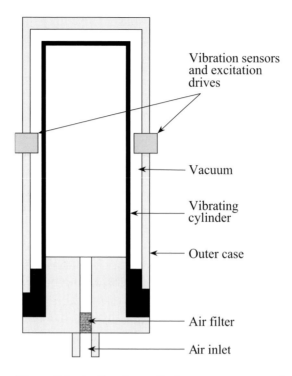

Figure 7.5. A vibrating-cylinder pressure sensor is about 10 cm long by 6 cm diameter. The inner, vibrating, thin-walled cylinder is open to atmospheric pressure on its inside while its outside is surrounded by a near vacuum in the narrow space between it and the outer cylinder. Inserted through the walls of the outer cylinder are three sensor units spaced at 120 degrees, which set the inner cylinder vibrating and also pick up the frequency of the vibration, giving a measure of barometric pressure.

The method is sufficiently accurate to measure barometric pressure to 0.1 hPa with a long-term drift of typically 0.05 hPa, making the sensor suitable as a transfer standard.

Siting of barometers

The majority of environmental instruments have to be deployed out of doors, but in the case of barometers there is no need for this, and they are usually installed in an office. It is necessary to protect barometers from large and sudden temperature changes, which can introduce errors through differential heating, and for this reason exposure to the sun through a window or to other heat sources such as radiators must be avoided.

On a windless day, pressures experienced indoors are correct, although the opening and closing of doors can induce quite large instantaneous pressure surges. More problematic is the effect of wind blowing against a building and through windows and doors and across openings such as chimneys. This must be avoided as it can induce quite large errors; many small openings are preferable to a few large ones.

Where wind-induced errors are a problem, for more precise work there is a device known as a *static pressure head* consisting of a vertical, hollow cylinder with a fin that turns it (wind-vane-like) into the wind. Holes, positioned around the cylinder relative to the wind flow, ensure that the internal pressure is independent of wind effects. The cylinder is connected to the aneroid barometer by a tube. Such a device may also be needed if the barometer is used in an air-conditioned room.

At AWS sites, the barometric pressure sensor, which is usually in a small box, will have to be housed in a hut or in the enclosure that houses the logging or telemetry equipment. The sensor might, therefore, be exposed to greater variations in temperature than one housed in an insulated building, and the temperature may change more quickly. However, electronic pressure sensors are much smaller and so not likely to suffer from temperature-differential problems. Wind might still be a problem and this needs to be considered, much depending on the nature of the housing.

References

Banfield, E. (1976) *Antique Barometers.* Baros Books, Trowbridge, UK. ISBN 0 948382 04 X.
Burton, B. J. (1993) Atmospheric pressure, and temperature aloft. *Weather*, **48**, 141–7.
McConnell, A. (1994) *Barometers.* Shire Publications, Princes Risborough, UK. ISBN 0 7478 0240 8.
Young, M. V. (1994) Back to basics: depressions and anticyclones: Part 1. *Weather*, **49**, 306–11.

8

Precipitation

The variable

A history of raingauging

The first written reference to rainfall measurement was made by Kautilya in India in his book *Arthasastra* in the fourth century BC (Shamasastry 1915). The next reference comes from the first century AD in *The Mishnah*, which records 400 years of Jewish cultural and religious activities in Palestine (Danby 1933). But neither the Indian nor Palestinian measurements continued for long. They were just isolated events, doomed to be ignored and discontinued. There were to be no more quantitative hydrological or meteorological measurements for another 1000 years – a period in which scholars believed, or were forced to believe, that one turned to the sacred scriptures for answers to questions such as 'Where do springs arise from?'. It was from China, around the year 1247, that the next known reference to quantitative rainfall measurement comes, and during the fifteenth century the practice of measuring rainfall was introduced into Korea, probably from China.

The first raingauge to be operated in Europe was made by the Italian Benedetto Castelli in 1639, a Benedictine monk and student of Galileo. Castelli measured rain only once, using a graduated glass cylinder about 12 cm in diameter and 23 cm deep. He does not seem to have considered doing this on a regular basis. In the 1660s Sir Christopher Wren made the first-known British gauge and later designed a second, which was probably the first ever that used a *tipping bucket* (Grew 1681). The first continuous record of rainfall was made by Richard Townley, in Lancashire (Townley 1694), from 1677 to 1703; the gauge consisted of a 12 inch (30 cm) funnel on the roof connected by a lead pipe into his house, the collected water being measured in a graduated cylinder.

Interest in measuring rainfall increased rapidly in the eighteenth century, worldwide, and so only the key points can be recorded here.

Dobson (1777) was amongst the first to expose a raingauge to today's standards, on a large grassy patch overlooking Liverpool. It was 12 inches (30 cm) in diameter and was part of a larger experiment concerned with evaporation. Most gauges, until 1770, had been exposed on roofs. Exposure on roofs has its problems, as William Heberden (1769) was the first to suspect. To investigate this, he operated two identical gauges, one on a chimney of his house and one in the garden, with a third gauge on a 150 foot (45 m) tower of Westminster Abbey. Readings were taken every month for a year and showed that the gauge on the chimney caught only 80% of that in the garden while the gauge on the tower caught only just over 50% (Reynolds 1965). He could not explain why this was, and it was 100 years before Jevons (1861) demonstrated that the reduction in catch with height was due to wind effects.

At about the same time George Symons, a young assistant in the newly formed Meteorological Department of the Board of Trade (the first UK Met. Office) became interested in rainfall and its measurement, fired by the drought years of 1854–8. It became a lifetime commitment and personal crusade. He amassed rainfall data from many sources and experimented with the various gauges then in use. In 1860 he published *English Rainfall*, which contained 168 annual totals, mostly from southern England but also from Guernsey and the Isle of Man (Symons 1864, 1866). He resigned his post at the Board of Trade in 1863 to devote all his time to the task of improving rainfall measurement and data collection, much to the debt of meteorology and financial difficulties for himself. By the time he died in 1900, he was receiving records from 3500 sites. Undoubtedly Symons was one of the most significant figures there has been in rainfall measurement and data collection.

Units and definitions

Precipitation includes rain, drizzle, snow and hail, but not condensation in the form of dew, fog, hoar frost or rime even though they can produce trace readings in a gauge of up to 0.1 mm. The total precipitation is the sum of all the liquid collected (including the water produced from melted solid precipitation), expressed as the depth it would cover on a flat surface assuming no losses due to evaporation, runoff or percolation into the ground. While inches have been used in the past to measure precipitation, and still are in some countries, millimetres are the more usual unit today (0.01 inches is about equivalent to 0.25 millimetres). The measurement of snowfall is treated separately later.

Factors affecting all conventional raingauges

The rain collector

Its diameter   In the 1860s, Colonel M. F. Ward did experiments in Wiltshire
to compare the performance of rain collectors of different diameters, from 1
inch to 2 feet (2.54 to 60 cm) as well as square funnels. Catch was found to be
independent of size, certainly above 4 inches. This was confirmed by tests by
the Revd Griffith, also in England, in the 1860s, during which he operated 42
types of gauge (Kurtyka 1953, Reynolds 1965). Although one set of tests in the
US suggested that small gauges caught more than larger ones, it is probably
safe to say that the area of the funnel does not matter provided it is bigger than
about 10 cm in diameter.

The use to which the gauge is to be put in part decides the best size of the
collector. For a daily manual reading, too little water makes for inaccuracy,
while too much needs a large container to hold it and makes measurement
inconvenient. The UK Meteorological Office at one time preferred an 8 inch
gauge but later changed to Symons's 5 inches, which is what is used today for
manual, daily observations. If the gauge is of the chart recording type, there
must be sufficient water to operate the float mechanism and move the pen
across the paper chart positively, while if it is a tipping bucket gauge there is a
lower limit below which it is inadvisable to go since a very small bucket tends
to have higher errors. Because the water measured by recording gauges is not
usually stored but allowed to run away, there is no collection problem and so
the bigger the collector, within reason, the better.

Constructional materials and evaporative loss   Traditionally, copper has been
used to make raingauge funnels (in the UK), although today anodised aluminium, stainless steel, galvanised iron, fibreglass, bronze or plastic are all used
(although not for the UK standard 5 inch gauge, which retains its copper
construction, for the present). What is important about the material is how
well it allows the water falling in it to run off and be collected or measured.
Materials do not all behave in the same way with respect to water.

Water first wets the surfaces of the funnel and the tubes leading to the bottle
or tipping bucket and only after this does it start to flow. When the rain stops,
the water left adhering to the surfaces does not drain into the bottle or bucket
but evaporates. In the case of a 5 inch UK manual gauge, the amount lost is
about 0.2 mm per rain event. There can also be some loss from within the
storage bottle, although this can be kept low by minimising ventilation, and by
housing the bottle below ground to keep it cool. Gratings within the funnel,
acting as filters to keep debris out, hold water which is lost by evaporation.

Any debris lying in the funnel will also hold water, which will thus be lost. If rain falls as intermittent light showers, loss by evaporation can amount to a significant proportion of the total catch; in the case of occasional heavy rain, however, the proportion will be much less.

Instructions given with regard to the UK 5 inch gauge is that it should not be polished (although it should be kept clean) – a matt, oxidised surface encourages runoff better than a shiny clean one. Essery & Wilcock (1991) observed a *settling-in period* of a year after new unweathered gauges had been installed, demonstrating that shiny new copper tends to suffer more from evaporative losses. At one time gauges were painted, but this is not advisable (on the inside of the funnel) because paints can cause the water to form large drops that stick, and can also absorb some water if they become porous.

Depth of the collector   One of the important factors in Symons's design is the depth of the collector, which is reflected in the UK Meteorological Office 5 inch gauge (see later), which has vertical sides 4.5 inches (11.4 cm) deep. This combination of diameter and depth prevents virtually any outsplash of large drops or the rebound of hailstones. Many gauges are to this pattern, but some have less deep sides above the funnel and this could result in loss of catch through outsplash. The problem with deep sides, or at least the resulting tall cylindrical shape of the gauge, is that this increases wind-induced errors (see below).

Levelling   It is not uncommon to see a raingauge leaning over at an angle. It may never have been installed correctly, or the ground may have settled after installation, or it may have been knocked by a lawnmower. Whatever the cause, the gauge will read wrongly. In windless conditions a tilted gauge will catch less rain, while if tilted into the wind it will catch more and if away from the wind less than it should. An error of about 1% occurs for each $1°$ of tilt (Kurtyka 1953).

Precision of collector dimensions   If the diameter of the collector is not as specified, its area will be wrong and the wrong amount of rain will be caught. For example, if the area is supposed to be 200.00 cm^2 (requiring a diameter of 15.95 cm) but the diameter is out by 0.5 mm, the catch will be in error by 1.25%. Distortion of the rim of the funnel, by knocks and dropping, will also cause errors.

Siting of gauges   Siting a raingauge near to a large object, such as a tree, is unwise. Nevertheless the number of gauges that are so exposed is surprisingly high. Perhaps the trees grow and are not noticed; perhaps there is nowhere else for the gauge. A raingauge should be installed no nearer to an object than four times the height of the latter. This applies not only to trees and buildings but also to other instruments such as temperature screens. Some sheltering may be beneficial, however, provided it is kept to the distance specified, because it lessens the adverse effects of wind.

In a forested area, one solution is to site the gauge in a clearing, provided the clearing is big enough to meet the above criteria of distance. If this is not possible, gauges are sometimes installed on a mast just above the forest canopy. If it is necessary to measure how much rain reaches the ground beneath the trees, the situation is more complex (see Chapter 6 on evaporation).

The effects of wind on raingauge catch

The effects of wind on raingauge performance cause by far the greatest errors of all. The raingauge acts as an obstruction to the wind, causing the wind to speed up over the top of the collector and around its sides, while eddies form within the funnel. Speeding up over the top causes some of the drops that should have been caught to pass over the funnel instead of falling into it, while eddies can lift drops out of the funnel that have successfully been caught (Fig. 8.1). These effects increase with wind speed and, since this increases with height, so too do the losses. In 1881 Symons demonstrated, through many experiments, that wind was indeed the cause of a reduced catch with height and showed that, if a 12 inch (30 cm) high gauge is taken as the reference, a gauge at only 2 inches (5 cm) above the ground caught an extra 5% while gauges at 5 and 20 feet (1.5 and 6 m) lost 5% and 10% respectively (Kurtyka 1953). These were averages and will vary from one rain event to another. In 1893 Abbé showed that the effects of wind depend on the texture of the rain, being more pronounced in drizzle than for larger drops. The problem is even greater for snow with its light flakes and larger area.

Wind shields   An early attempt to combat the problem was made in the USA by Nipher (1878), who surrounded the gauge with an inverted trumpet-shaped *shield* (Fig. 8.2) that deflected the wind downwards. Such a shield can, however, become filled with snow, overcapping the gauge, and so, in 1910, Billwiller (in Germany) cut the bottom out of the Nipher shield to allow snow to pass through. He also removed the outer horizontal collar, although this was later shown to be an important part of the Nipher design. Also, in the USA, Alter (1937) designed a shield consisting of loosely hanging

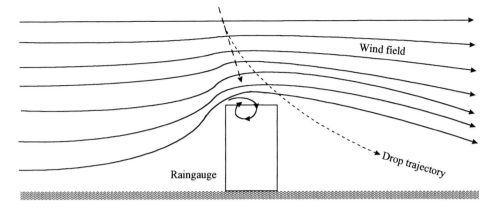

Figure 8.1. The airflow over a cylindrical raingauge is distorted by the presence of the gauge; it speeds up over the top, carrying away drops that should have been caught. Increased eddies within the collector also lift out drops and carry them away – a large rim on a collector increases this effect.

strips of metal (Fig. 8.3) which had the property of helping to prevent snow accumulation. It is, however, large and somewhat expensive.

The turf wall   In the UK, Huddleston (1933) designed his *turf wall* (Fig. 8.4), as part of a series of experiments from 1926 to 1933 near Penrith, the object being to find the best raingauge exposure for mountainous and windy conditions. He also tested fences and sunken gauges but concluded that the turf wall was best and then went on to determine the optimal design for it.

The pit gauge   Stevenson (1842), in the UK, was probably the first to do anything to combat the wind problem, and his design, with modifications, is still the best. He buried the raingauge so that its rim was just above ground level and surrounded it with a mat of bristles to prevent insplash. In this way the gauge did not interfere with the flow of the wind at all.

Later, in Germany, Koschmieder (1934) designed a modern *pit gauge* (Fig. 8.5) and this has been shown to be the best way of exposing a raingauge to minimise wind effects (Rodda 1967a, b). By keeping the gauge out of the wind flow, all wind effects are avoided. The grating around it minimises eddies within the pit. It has its problems, however, since it can become snow- or sand-filled (as can the turf wall) or flooded, while at unattended sites it can become overgrown with vegetation. Nevertheless a pit gauge is the best way of exposing any raingauge, as set out in several WMO reports (Sevruk & Hamon 1984), and is currently the topic of a European Standards Working Group on raingauges.

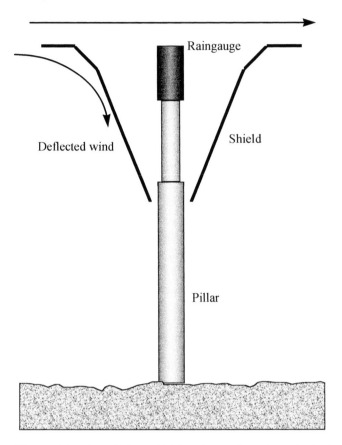

Figure 8.2. The Nipher shield reduces the effect of wind on the catch of a gauge by deflecting the wind downwards, but it tends also to collect snow. The Billwiller screen is similar but has the bottom and the flat collar removed to inhibit snow accumulation.

Aerodynamic gauges   Another way of combating the effects of wind on raingauge catch is to design the profile of the gauge so as to present as little interference to the air flow as possible. Most gauges are cylindrical in cross-section and these cause the maximum of interference to the passage of the wind. Robinson & Rodda (1969) tested several raingauge designs in a wind tunnel, observing the flow of air with smoke trails. Field tests were also carried out, comparing the catch of the gauges, which included a funnel-shaped gauge of 5 inches (12.7 cm) diameter with a half-angle of 45°, the funnel gauge catching most of the rain except when it was heavy, when outsplash was suspected of causing losses.

Adopting a mathematical modelling approach, Folland (1988) analysed the flow of air over a cylinder and proposed a first-guess new design, which

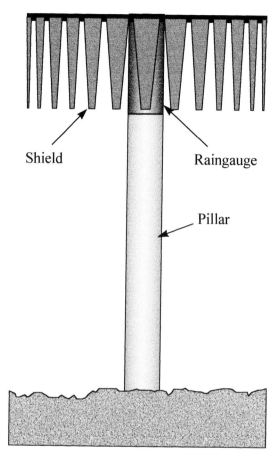

Figure 8.3. The Alter shield was designed to avoid the problem of snow accumulation; its individual leaves swing freely, preventing the build-up of snow. In strong winds the leaves on the windward side swing in to form a shape similar to the Nipher shield, deflecting the wind downwards. The Tretyakov shield has the Nipher profile but is made of individual (fixed) leaves similar to the swinging leaves of an Alter shield.

he called the *flat champagne-glass* raingauge, a combination of a cone with a diameter of about 25 cm and a half angle of 35°, with near-vertical sides to minimise outsplash. Strangeways (1996) made and tested several collectors to this design (Fig. 8.6) and carried out comparative field tests on them demonstrating that the catch was indeed improved, but that in very heavy rain outsplash appeared to be the cause of a slight fall-off in performance. A vertical partition within the gauge, in the form of a cross, however, helps to lessen outsplash while also minimising internal eddies.

While Folland was undertaking his theoretical work, Strangeways (1984) was also experimenting, independently, using the findings of Rodda and

Figure 8.4. The turf wall, seen here at the Institute of Hydrology in Wallingford, was invented by Huddlestone in 1933, its purpose being to shield the gauge from wind. In this particular case it has a Mark 2 five inch manual gauge at its centre. But, like the Nipher shield, it can collect snow.

Figure 8.5. The best technique for exposing any raingauge, in order to prevent wind effects, is to instal it in a pit surrounded by a grating. The construction seen here conforms to WMO specifications. But like the turf wall it is liable to fill with snow, sand and vegetation, if left unattended.

Robinson. The designs that emerged from these developments were similar to those of Folland's. Field tests on two such designs (Hughes *et al.* 1993) confirmed that an improvement in catch was obtained with cone-shaped collectors, but that they did not give results as good as those from pit gauges. Aerodynamic designs have also been pursued by others, such as Wiesinger (1996) in Austria.

Figure 8.6. By improving the aerodynamics of raingauge design, wind effects can be lessened, although not quite as effectively as by exposure in a pit. Nevertheless, pit exposure is not always possible; then a design such as that illustrated here becomes a good second best.

Correction procedures, field intercomparison and wind-tunnel tests According-ing to Sevruk & Klemm (1989) there are over 50 different types of raingauge in common use worldwide, all differing in size, shape, material, colour and the height at which they are exposed above the ground (from 0.2 to 2.0 metres). The number is greater if the newer automatic gauges are included. Because each gauge has its own characteristic errors, accurate intercom-parison of rainfall data from around the world is not possible. To deal with this problem, the WMO organised field intercomparisons of the major na-tional gauges of the world, with pit gauges as the reference, at 60 sites in 22 countries (Sevruk & Hamon 1984). However, such intercomparisons lump many rain events together over a period and can only give a rough correc-tion. Wind speed and raindrop size are the two key factors in how different raingauges behave; if measurements of these are also available it becomes possible to correct readings with higher precision.

As mentioned above further improvement can also be obtained by testing gauges in a wind tunnel, in which variables can be manipulated and investi-gations more precisely targeted at individual factors. Measurements of the speed-up of wind over collectors have shown that it is particularly depend-ent on the size and shape of the orifice rim: the smaller the rim, the smaller the increase (Sevruk *et al.* 1991).

Sevruk (1996) has gone on to develop mathematical simulation techniques of the wind-induced error, using a three-dimensional numerical procedure. First the air flow around the gauge is simulated and then the trajectories of different drop sizes are computed in the simulated flow fields. The result is a formula giving the wind-induced error as a function of wind speed, rate of rainfall and drop size, for each raingauge. This is relatively new work and has the advantage that the corrections can be derived at low cost in the computer without the need for field or wind-tunnel tests. However, to use the method to make corrections, it is necessary to know the wind speed at gauge height and the drop size, which adds to the complexity of the instruments, and is usually not available for past data, although it can sometimes be inferred.

From point measurements to areal estimates

The purpose of measuring rain (and indeed most variables) is usually to estimate areal averages rather than to obtain a single-point observation. Over anything other than a very small area, several gauges are usually necessary to obtain a reasonable average. While some specialised investigations may use a high density of gauges, even the best meteorological network rarely has more than one gauge for every 25 km^2 (25 000 000 m^2). With a funnel diameter of 12.7 cm (5 inches) the area is 126.7 cm^2 (or 0.01267 m^2). This means that the area from which rain is collected is around 1 in 2 $\times$ 10^9 of the area over which rainfall is to be estimated. In most parts of the world the raingauge density is much smaller than this and over the seas almost non-existent. While some argue that in a situation like this it is not worth measuring rain very carefully, the contrary is the case, for without accurate point measurements, the spatial variability of rainfall cannot be established because it will be swamped by raingauge erors. Also, a rainguage with a systematic error (such as undercatch through wind effects) will always give a bias to areal estimates. It is all the more important, therefore, to try to get the point readings as accurate as possible, through careful choice of site, careful positioning of the gauge at the chosen site and use of the best raingauge possible.

Having done everything possible to get good point measurements, how can these be converted into areal estimates? Imagine an area completely covered in square collecting funnels. The area's average rainfall for any period would be the average reading of all the individual gauges. But in practice most of these gauges will be missing. So how is it possible to estimate the average from just a few gauges?

It would be possible to take the average reading of them all. But there might be several gauges clustered together, and they might be in a region that is not typical of the whole area. A technique often used is to work in percentages rather than actual millimetres of rain, each raingauge reading being converted into a percentage of the *gauge's* average annual catch, determined over a period. This percentage is then converted back to rainfall amounts using the *catchment* annual average. While this produces more uniform readings, the method confers on each raingauge the same importance, even though some may be clustered together, or perhaps be in an unrepresentative part of the area. One approach to this problem is to give each of the gauges a weighting that takes account of the relative importance of its contribution to the average (Jones 1983). A widely used technique for doing this is the *polygon method* (Thiessen 1911) in which the area is divided into polygons such that the polygon around each gauge is that part of the area nearer to that gauge than to any other. The weighting for each gauge is then the ratio of the area of its polygon to the total area. But the gauge surrounded by the largest polygon may not be at a representative site.

An alternative to the polygon is to derive the gauge's weighting by dissecting the area into triangles. Briefly the procedure is to construct a mesh over a map of the area and for each mesh point to search for a triangle of raingauges that surround it, but limited to a certain distance. If no such triangle can be found the nearest three gauges are selected. How the mesh is constructed, the weights determined and the maximum distance is selected, is complex (Jones 1983). Other methods calculate the weightings of each gauge based on criteria such as the angles each subtends at the major and minor axes of the area.

Where a network of raingauges is to be installed from scratch, the opportunity exists to choose the sites. How can this best be done? Gauges could be placed on a regular grid, but this might mean that some are in very unsuitable places such as on a hilltop. In practice it is more realistic to site gauges where there is easy access, or where an observer lives (in the case of manual gauges). But a choice made in this way will probably be biased, the remoter, higher and least accessible all being omitted. The area could first be saturated with gauges, to get somewhere near the ideal of covering the whole area with gauges, irrespective of the difficulty of getting to them. Then, when there has been a chance to study their relative catches, the number of gauges could be reduced, keeping a watch on how this reduction affects the accuracy of the mean areal rainfall. Sometimes gauges turn out to be key ones, being very representative of a large area. Another technique is to identify topographic *domains* (Rodda 1962), in which factors such as the slope, aspect and elevation extend over a clearly definable distance, a gauge being placed at random within the domains; but this is not realistic in flat areas or where the landscape is rugged. There is no ideal solution, and the best method for any one situation needs to be selected with an informed knowledge of the options. A combination of two or more of the above methods may be appropriate.

Manual gauges

As Sevruk & Klemm (1989) have pointed out, there are over 50 different types of daily-read manual gauge in general use worldwide, but as most gauges fall into clearly defined types it is only necessary to describe typical examples of each, and the UK case will be used for this.

The classic five inch manual gauge

The UK Meteorological Office Mark 2 gauge demonstrates the general features of most manual gauges and is based on Symons's 5 inch model (12.70 cm diameter, with an area of 126.7 cm^2), which has now been standard for most of the twentieth century in the UK (Fig. 8.7) and will continue in use into the twenty-first century.

A copper funnel with an accurately turned bevelled brass rim and deep sides fits onto a copper base, set into the ground, with a splayed-out base to give it more stability. For interchangeability, the cylindrical parts are made of drawn copper tube (not of rolled, soldered sheet). Inside the case is a removable copper container and inside that is a collecting bottle. The gauge is installed so that its rim is 12 inches (30 cm) above the ground, which should be of short-cut grass or failing this, gravel. A hard smooth surface should not be used, to avoid insplash. The bottle holds the equivalent of 75 mm (3 inches) of rain. The removable copper container provides extra capacity in case the bottle overflows in heavy rain or the gauge is left too long, together holding 140 mm (5.5 inches). A cheaper model, known as the Snowdon pattern, has straight sides and so is less stable. A still cheaper version is made in galvanised steel. Another cheaper, but larger, gauge is also available with a funnel aperture of 20.32 cm (8 inches) and a total measuring capacity of 180 mm of rain (7 inches). In the USA, 8 inches is most commonly used.

The gauges are read daily using a graduated glass cylinder into which the collected rain is poured. The cylinders are either flat-based or tapered, the latter decreasing in diameter from 1 mm of rain down, for greater accuracy when small amounts of water are collected (Fig. 8.7). They are marked in 0.1 mm steps with an additional mark at 0.05 mm, figuring being at 0.1 mm, 0.5 mm and every 1 mm between 1 mm and 10 mm. The maximum error for the measure is ± 0.02 mm up to 2 mm and ± 0.05 mm above 2 mm. In reading the amount of water in the cylinder, the procedure is to note the reading closest to the bottom of the water meniscus to the nearest 0.1 mm. Weighing the bottle and water and subtracting the weight of the dry bottle gives a more accurate measure, if needed.

Figure 8.7. The 5 inch, Mark 2 raingauge preferred by the UK Meteorological Office. Rainwater collected by the funnel is stored in a bottle within the case, which extends below ground, and is measured daily in a graduated cylinder of the type shown. The base of the measuring cylinder tapers to allow small volumes of water to be measured more precisely.

When gauges cannot be read daily, a larger container is necessary so that weekly or monthly amounts of rain can be stored. In the UK two such models are used, both having a funnel diameter of 12.70 cm, one holding 680 mm of rain (26.8 inches) and a larger model holding 1270 mm (50 inches), for areas of high rainfall or where the gauge may have to be left for two months. It is known as the Octapent gauge because it is a merging of a 5 inch funnel with the base for an 8 inch funnel. It has a splayed base similar to the Mark 2 gauge. These two gauges have a precisely fitting removable inner container to hold the water, the water first being measured roughly in the container by means of a dipstick to get a rough estimate of the amount; thereafter it is measured more precisely using a graduated cylinder. Provision is made to insert a flexible tube that will collapse under the pressure of freezing. A modified version of the Snowdon gauge, known as the Bradford gauge, is deeper and can hold 680 mm of rain; it was designed for use in similar circumstances to the Octapent.

Other types of manual raingauge

There are more than 150 000 manually read gauges in use throughout the world. Sevruk & Klem (1989) presented an analysis of their distribution and type and showed that the most widely used (in 1989) was the German Hellmann gauge, with 30 080 in use in 30 countries. The Chinese gauge is second with 19 676 used in three countries, and the English Mark 2 and Snowdon third, with 17 856 operated in 29 countries, all three types totalling 67 612 and accounting for about half the world's raingauges. The Hellmann gauge has a funnel of 200 cm^2 or 15.96 cm diameter (6.28 inches) while the Chinese gauge has a collecting area of 314 cm^2 (diameter 20 cm or 7.9 inches). Like the Hellmann gauge, the Chinese gauge is made of galvanised iron.

The remainder of the most commonly used world's gauges are as follows.

> Russia. 13 620 gauges in seven countries, 200 cm^2, galvanised iron
> The USA. 11 342 gauges in six countries, 324 cm^2 (8 in), copper
> India. 10 975 gauges just in India, 200 cm^2, fibreglass
> Australia. 7639 gauges in three countries, 324 cm^2 (8 in), galvanised iron
> Brazil. 6950 gauges just in Brazil, 400 cm^2, stainless steel
> France. 4876 gauges in 23 countries, 400 cm^2, galvanised iron or fibreglass
> Totalling 55 402

The grand total of all these is 123 014, although Sevruk estimates there are about 150 000 gauges in all worldwide.

No doubt this has changed since the count was done and it probably missed gauges used by small organisations, but it gives a good idea of the situation. It does not include automatic gauges of any type, just manual ones.

Mechanical, chart-recording gauges

Recording raingauges are used mostly to supply information on the times when rain starts and stops and to give an approximate indication of the rate of rainfall. It is usual to operate a manual gauge nearby to act as a reference.

Mechanical recording raingauges are of two main types – those that cause a pen to move across a paper chart through the movement of a float and those that use a balance.

Float-operated recorders

The rainwater collects in a container in which there is a float that moves upwards as the container fills, so that a pen also moves upwards across a paper chart turned by a spring-driven clock. The container is of smaller diameter than the collecting funnel, giving magnification to the pen movement. In early models, the container had to be emptied by hand when it was full, while in later designs mechanisms of various types were devised to make the pen fall back down to the bottom of the paper chart and restart its climb back up. It could do this several times before the container was full, but it still then needed emptying by hand. The gauges used today all have methods of emptying the container automatically each time it becomes full. How the container empties is the main difference between recorders, but in fact all current models empty by siphoning.

The tilting siphon gauge   The most popular chart-recording raingauge in the UK, the tilting siphon gauge, was designed by Dines in 1920, the rain collected by a funnel being fed into a cylinder containing a float (Figs. 8.8(*a*), (*b*)). The float rises as water enters the cylinder, moving a pen up a paper chart. When the float approaches the top of the cylinder it releases a catch that causes the cylinder, mounted on knife edges, to tip over to one side. The action of tilting causes the siphon tube to be suddenly flooded with water, kick-starting the siphoning process – which must start positively so as to prevent 'dribbling'. When nearly empty, and while still siphoning, the cylinder tips back to its vertical position (for it is counterbalanced) ready to repeat the process. Because it takes time to siphon the water out, the mechanism is designed to save the incoming water during the siphoning process.

Two models are made, one for temperate regions having a collector of diameter 28.73 cm (area 648.4 cm^2), which siphons for every 5 mm of rain, taking from 6 to 8 seconds to empty. The model for tropical, heavier rain is the same except that the funnel has only one-fifth the area of the temperate model and so siphons every 25 mm (taking the same 6 to 8 seconds, for the internal mechanism is the same). The tropical model is surrounded by a white outer case to shield it from the sun. Charts can be for daily, weekly or monthly records.

The natural siphon gauge   The natural siphon recorder, designed by Negretti and Zambra, is very similar to the tilting siphon recorder and predates it by 20 years. The siphon consists of two coaxial tubes to the side of the float cylinder, constructed so that the annular gap between the inner and outer tubes is small,

(a)

(b)

Figure 8.8. (*a*) A Dines tilting siphon recording raingauge. The paper chart mechanism, visible through the window of the gauge, records the movement of the pen. (*b*) Water enters the container on the right (through the small elongated funnel); as it accumulates it raises a float, connected directly to an arm that terminates in a pen in contact with the paper chart. The container is balanced on a knife edge and when full the float releases a catch that allows the container to tilt, starting siphoning abruptly. The counterbalance (bottom right) brings the cylinder back to its upright position when siphoning is almost complete.

causing siphoning to start abruptly without any special mechanical trigger being necessary. When the water reaches the top of the outer tube, capillary action ensures that siphoning starts decisively, pushing all the air out and down the inner tube. Conversely, after siphoning, and once air gets to the top of the tube, siphoning abruptly stops. The simple construction of this instrument means that it is cheaper and there is less to go wrong than there is with the tilting syphon gauge; however, there is no means of saving the water that enters the funnel during siphoning.

As with the tilting siphon, there are two models: the temperate-climate design has a funnel of 8 inches diameter (20.3 cm, area 324 cm^2) and siphons for each 10 mm of rain, taking from 12 to 15 seconds to empty. The tropical model is the same but has a funnel of area 129.7 cm^2, that is, 2.5 times smaller, siphoning every 25 mm of rain.

The Hellmann siphon gauge   Hellmann designed a float gauge in Germany in 1897 using a siphon with a long tube that ensured accurate action. It has a collecting area of 200 cm^2. It became widely used in central Europe and, along with the Dines and Negretti recording gauges, probably accounts for most of this style of gauge in use today. As it is very similar to the natural siphon gauge discussed above, it needs no further description. Having a large container at the bottom to store the water after measurement (as back-up), it is bigger, standing 1.2 m high.

Weight-operated recording gauges

Weighing raingauges operate by recording the total weight of precipitation as it accumulates in a container, either by suspending the container on a spring or on the arm of a balance (Fig. 8.9). In both designs the weight of the water forces the container downwards, the vertical movement being magnified by lever linkages to move a pen. In some designs, a dual-traverse action is used so that the pen records half of the gauge's total capacity in the upward direction, the second half downwards. To prevent wind shaking the pen, and so introducing movement into the trace, the pen is damped by an oil dashpot.

Osler made one of the first weighing raingauges, in the UK in 1837, in which the container was counterbalanced by a weight on the opposite side of the fulcrum. A rod, connected to the weighted side of the balance, moved a pencil over a paper chart. The receiver had a siphon that emptied every 0.25 inches of rain. Two dozen or so different types of weighing recorder have been made in the period since 1837, some being emptied by hand, some siphoning the water out, while others used containers that tipped over when full, so emptying themselves.

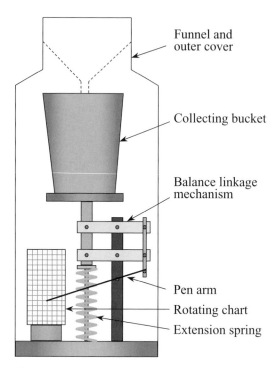

Figure 8.9. An alternative to float recorders is the weighing raingauge, illustrated here schematically to show the principle. This type can have the funnel removed to collect snow.

The advantage of periodic emptying is not just that operation can be continuous but that the balance mechanism can be made more accurate because it does not have to work over a wide range of weight. For example, in the case of an 8 inch diameter collector (area $324\,\text{cm}^2$), the weight of water representing 0.1 mm of rain is 3.24 grams. If the gauge does not empty itself but instead has the capacity to collect up to, say, 12 inches (30.48 cm) of rain, the balance must be able to measure changes of 3.24 g in a total weight of 9884 g of water, or 0.03% of full scale. In reality only about 0.5% accuracy is achievable in a mechanical field balance, representing little better than a measurement of rain to 1 mm. In the case of float-operated recorders the same problem exists, but all modern float recorders siphon periodically, in such a way that the recorder needs to detect changes of 0.1 mm over a range of 5 mm, a resolution of only 1 in 50 (in the temperate models), or 0.1 mm over 25 mm, a resolution of 1 in 250 or 0.4% (in the tropical models). For anything other than approximate measurements, therefore, a weighing gauge, like a float gauge, must empty itself regularly.

The containers of weighing raingauges tend to be more open than in the case of float gauges, many using an ordinary galvanised bucket (pail), making them more vulnerable to evaporative loss, although this can be prevented by adding oil to the container so that it floats on the water.

Rate-of-rainfall recorders

There are two main types of raingauge that measure the intensity of rain – the *fixed volume* and the *calibrated orifice* (or *controlled orifice*) gauges. While they may be classed as rate recorders they can, with suitable integrators, also measure rainfall totals. The fixed volume type is more familiar as the tipping bucket (see below). The calibrated orifice type is much less common.

Calibrated orifice rate-of-rainfall sensors   There are three categories of this type of gauge. In the first, drops are formed that are then counted electrically, examples being gauges developed by W. Gallenkamp in Germany in 1887 and by W. J. Binnie in the UK in 1892; there are other more recent examples such as the gauge developed by the Radio and Space Research Station in the UK (Norbury & White 1971). Drop size is approximately proportional to the product of surface tension and the tube's bore (in the case sited, 2.8 mm diameter produces drops of 0.07 ml). Norbury and White state that at a constant flow rate, the drop volume varies by up to 3%, although at high rainfall rates by up to 10%. Also, as drop size is affected by changes in surface tension and as this varies with the degree of pollution in the water and its temperature, additional errors occur. For the same gauge, over a temperature range of 5–35 °C, a change in drop size of up to 5% can occur. Clearly errors of this magnitude would be serious if precise totals of rain were required. However, for relative rather than absolute intensities, it is an attractively simple method with no moving parts.

In the second type of calibrated orifice gauge, the rainwater enters a container, from which it is also allowed to leave. The depth which the water reaches in the container is a combined function of the rates of input and output of the water. The input is controlled by the rainfall rate, the output is variable and controlled by a valve. In a gauge designed by Sprung in Germany in the early twentieth century, a weight-controlled valve over an orifice moves vertically as a function of rainfall rate; the higher the rate, the higher the valve and the more water is passed, the movement of the weight being recorded on a paper chart. The Spanish Jardi gauge of 1921 works similarly but the valve is controlled by a rising float.

In the third calibrated orifice type, the container has holes or a slit in its side, the amount of water retained being proportional to the rate of its input. The container is weighed, in the designs of F. J. Cornick in the USA and C. Nell in the UK in 1950, but the depth could equally well be measured by a float.

Nevertheless, rate-of-rainfall sensors, apart from tipping buckets in electronic systems, are comparatively rare, rather specialised and little used.

Electronic raingauges

Although there are exceptions, which will be explained later, most electrical raingauges work on the principle of collecting the rainwater in a funnel and measuring it in some way.

Tipping bucket raingauges

Despite their problems, tipping bucket gauges are by far the most commonly used type of automatic raingauge in the world today.

Basic principles   Sir Christopher Wren's design (mentioned in the opening chapter) was single sided – it filled, tipped over and fell back again. Crossley, in 1829, was the first to use a double-sided tipping bucket. Other designs followed throughout the nineteenth century in Europe and America and they have proliferated in the twentieth century. Less common has been the use of a water wheel, the principle being that instead of a tipping bucket a wheel rotates in discrete steps. But they are not used today.

Modern designs are all symmetrical (Fig. 8.10), although the shapes of the buckets vary from make to make. The size of bucket is important, for it is undesirable to attempt to make a bucket tip for very small amounts of water because the errors (described below) increase. Ten millilitres or more is an advisable minimum, and so to measure 0.2 mm of rainfall reliably requires a funnel of area at least 500 cm². Gauges are made with buckets that tip at intervals of 0.1, 0.2, 0.25, 0.5 and 1 mm. Which interval is chosen depends on the use of the data and on the conditions; the move today is toward 0.2 or 0.25 mm buckets.

The tip of the gauge normally moves a permanent magnet, fixed to an arm on the bucket, past a magnetic reed switch, giving a brief contact closure. Reed switches can suffer from 'bounce' but by ensuring that the counter circuits do not have too fast a response time, multiple counting can be prevented. Mercury-wetted switches also help to prevent this, as well as to reduce the other reed-switch problem, contact wear.

Invariably today a tipping bucket raingauge will be used in conjunction with a data logger, either built into the gauge as a single-channel recorder or as part of a multisensor system such as an AWS, which will include a multichannel

Figure 8.10. The most usual way to measure the water collected by a rain-gauge electrically is by means of a tipping bucket. In this figure, it has been removed from the gauge for clarity. With a collector of $500\,cm^2$ a bucket of this size will tip for each 0.2 mm of rain (10 ml of water). This particular design has a sloping central divider that prevents water from collecting on the 'wrong' side of the pivot. As it tips, a magnet in each of the arms attached to the bucket closes a reed switch in one of the fixed arms of the pivot support (to allow two recording devices to be used with the gauge).

logger. The contact closures of a gauge can be logged either as a total number over a period or as the time and date of each tip, giving an indication of intensity and time distributions.

Strengths and weaknesses The advantages of the tipping bucket are that it is simple, there is little to go wrong, it consumes no power and can be used with a variety of recorders. Its weaknesses are relatively few, but they can be a problem (Hanna 1995), as follows.

The record provided by a tipping bucket is not continuous, although if the tips are every 0.1 mm of rain this is close to being so. A UK Meteorological Office design overcomes this by weighing the bucket as well as counting its tips, giving an indication of the increasing amount of water on a continuous basis between tips (Whittaker 1991, Hewston & Sweet 1989). It operates by suspending the bucket on wires, the natural resonant frequency of which increases as the weight increases; the signal is picked off magnetically.

When the bucket has received the correct amount of water and starts to tip, water continues to enter the full side until it is in the level position, this rain being in excess and thus lost, although the error is only small. However, if the gauge is calibrated by passing water into it at a controlled rate, equivalent, say, to 10 or 20 mm of rain an hour, this error can be calculated for different input rates and automatic correction made for it during data processing (Calder & Kidd 1978). Calibration in this way is preferable, in any case, to using a burette to pass water into the bucket until it tips once, since an average calibration for 50 tips will be more precise; drop-by-drop calibration is also difficult to do precisely. A carefully measured volume, of around 1 litre, is passed slowly into the gauge and the number of tips counted over a period of perhaps 60 minutes, the bucket being set to tip for the required amount by adjusting the pillars on which it rests. Since this can be a long process – of repeated adjustment and test – it can be preferable to set the bucket close to the desired sensitivity, of say 0.2 mm, and to measure exactly what it is by the passing of 1 litre of water. Thus a gauge may tip for each 0.19 mm, this being used as a bucket calibration factor.

Some water inevitably sticks to the bucket after it tips, this being lost if the bucket does not tip again until after it has evaporated. If it does tip again, slightly more water will be required to tip the second time because the wet side is now slightly heavier, but on third and subsequent tips, balance is restored. If the bucket is calibrated by passing a litre of water through the gauge, the wetting effect is automatically compensated for and the error only occurs in the event of one isolated tip. The bucket's surface will change with use, generally accumulating a thin film of oxide, if made of metal, or of adhering dust, and the amount of water sticking may thus change. There is, therefore, a case for calibration checks some months after installation. This is one reason why tipping buckets should be as large as possible, since in the case of small buckets the amount of water adhering is a larger percentage of their total capacity and so the error will be greater. To minimise the adherence of water, some tipping buckets have drip wires at the point where the water leaves the bucket, although this only reduces the problem at that one point.

Because rain rarely ceases just as a bucket tips, some water is usually left standing in the bucket until it next rains. If this is within a few hours, there is not much of a problem, but, if rain does not occur again for some time, all the water left standing in the bucket may evaporate. While this can amount to one full tip, on average over a long period it will average out at half a tip.

Depending on the type of rain experienced this error can be large or small, light showers spaced by a day or so producing larger errors than long periods of heavy rain. This is another reason for having buckets that tip for small increments of rain (say 0.1 rather than 0.5 mm), for less will then be lost by evaporation.

Since the bucket must tip in response to a very small change in the amount of water in it, friction in the pivot or any variation of friction over time will affect performance. This is yet another reason for using as large a collector as possible. The type of bearing varies from make to make, including ball races, knife edges, rolling systems and wires under tension. Each performs differently.

Other types of electronic raingauge

Weighing gauges   By replacing the pen of the mechanical weighing raingauge by a mechanism that turns a potentiometer, an electrical output is obtained that can be logged. In electronic versions, the water is weighed by a straingauge load cell. Because electronic weighing is more precise than can be achieved in the field with mechanical springs and levers, greater accuracy is possible. One commercial raingauge senses increments of 0.01 mm of rain in a total of 250 mm – the capacity of its non-emptying container. Another model has a capacity equivalent to 1000 mm of rain.

An advantage of the weighing gauge is its ability to measure the weight on a frequent basis (just during rainfall), allowing the intensity of the rain to be measured in finer detail than by a tipping bucket. But for long-term unattended operation, some means of emptying the weighed container automatically is necessary, such as by siphoning.

Optical gauges   Optical raingauges have evolved largely as a result of spin-off from visibility-measurement developments, raindrops and snowflakes being detected by their effect on a horizontal beam of infrared light about one metre long. An infrared receiver detects the passage of the particles through the beam, either by being in line with the transmitter or by detecting backscatter or sidescatter from the precipitation particles, the resultant scintillation waveform being analysed. It is essentially a rate-of-precipitation sensor although, by integration, hourly and daily means of intensity can be obtained, thus giving the total amount of precipitation. The type of precipitation, whether rain, snow, a mix of the two or hail can be detected by analysing the spectrum of the scintillation signal. This sensor also gives a yes/no indication of the precipitation state, making it a useful instrument at, say, an airport or an

Antarctic base, where an observer can watch a screen and know whether it is raining or snowing and if so how intensely (another name for the instrument is a *present weather detector*). This can be particularly useful at night. It is also useful in cold regions where a large proportion of the precipitation falls as snow, which is difficult or impossible to measure with normal raingauges (see later), although the need to keep the optics ice-free means that power requirements might be high.

For indicating instantaneous precipitation rates, an optical raingauge has advantages over the tipping bucket but, for the long-term collection of precipitation amounts, its high cost (about £7000 to £8000 currently) makes the tipping bucket raingauge, at one twentieth the cost, inevitably preferable. Manufacturers of optical gauges (there are very few) quote an accuracy of from $\pm 1\%$ to $\pm 10\%$ (of actual intensity) depending on intensity.

An exception to the cost of optical raingauges being a deterrent to their use might be their relative lack of sensitivity to non-stable deployment, which could be an advantage on ships or buoys.

Precipitation detectors   Precipitation detectors sense whether rain or snow is falling, but not its intensity. Another name for them is *ombroscopes*. A small sensing surface of somewhere between 10 and 50 cm^2 inclined at an angle (to allow water to run off) detects the presence of water on it by measuring either the change in resistance, or the change in capacitance, between interleaving electrodes, owing to the presence of water. The plate is heated if the temperature falls below $+5\,^\circ$C, to detect snow. It is also heated when there is water on it, so as to dry it off quickly after rainfall or snowfall stops. Lower-power heating prevents fog and dew from forming (and from being interpreted as rain). Ombroscopes have a shorter history than raingauges, having been made only since the start of the twentieth century. They find use particularly in operational situations where the occurrence of precipitation needs to be known in real time or where its duration needs to be known but quantity is not required.

Impact-sizing sensors   Concern over drop size was mostly driven by an interest in cloud physics in the mid twentieth century, but a few drop-size detectors were also developed for use on the ground to measure the energy contained in the impact of each raindrop as it hit a surface, as a means of measuring rainfall quantity. Nevertheless, the main application was directed more towards measurement of the size-spectrum of raindrops. A few instruments have, however, been developed to sense the size and number of the individual raindrops, as they hit a surface, and thus the rainfall quantity, but they have not found wide use.

One particular type was based on the measurement of small changes in the value of a capacitor made of interleaving electrodes (very similar to those of a precipitation detector), covered by a thin glass plate to protect them. This type of sensor detects the volume of each drop, the changes in capacitance that occur as each drop lands and spreads out on the glass modulating the frequency of an oscillator circuit (Genrich 1989). It is the sudden arrival of the drop that is detected, the water remaining on the plate afterwards presenting a DC component that can be ignored. But it is not a direct method and it is necessary to find a reliable relationship between the change in capacitance and the volume of the drop that produces it. It is unlikely that this can ever be as precise as measuring the volume of the accumulated drops collected by funnel. Genrich suggested that such sensors might find application in conjunction with a tipping bucket gauge, the impact-sizing sensor being able to detect small amounts of precipitation better than a conventional gauge, while the tipping bucket is better at all other rates.

Another type of impact sensor attempts to measure the energy in the impacting drop as it strikes a sensing membrane; the energy depends on the drop's volume (weight) and its speed of impact, larger drops falling faster than small ones. But again finding a convincing conversion factor from impact intensity to volume of water is not simple. The speed of impact is also affected by the speed of the wind, as is the angle of fall of the drop. Devices such as this may have a small place in applications requiring some knowledge of drop-size distribution, but they do not replace conventional raingauges.

Again, as with the optical raingauge, applications for impact sensors might be found at sea where conventional gauges are problematic. Indeed this is being investigated, although in relation to the sound-signature of the bubbles that raindrops produce in the water rather than the impact of the actual drops themselves (Chapter 13).

The measurement of snow

Snow is very difficult to measure, as will become clear.

Measuring snow as it falls

Wind effects   Snow is more easily blown by the wind than is rain, and this greatly worsens the aerodynamic problems. Above a wind speed of a few metres per second, the use of raingauges for measuring snow is not feasible without some form of screen. Pit gauges are obviously not practicable, although a possible alternative is the use of aerodynamically shaped collectors such as that shown in Fig. 8.6. After collection, wind eddies can lift out dry snow that has been successfully caught. This complication can be reduced by removal of the funnel, so presenting a deeper container, which lessens the reach of the eddies, or by introducing a cross-shaped vertical divider into the funnel.

As the wind speed increases so does the angle of fall of the snowflakes, becoming increasingly more horizontal; this causes the flakes to approach the gauge at such a small angle that the orifice presents too thin an ellipse for them to enter. After falling, dry snow can also be picked up by the wind and carried elsewhere as spindrift (Fig. 8.11), which if caught by gauges cannot be differentiated from true snowfall.

Measuring snow with manual raingauges   If the snowfall does not overcap or bury the gauge, and the wind is light, it is possible to use conventional manual raingauges to measure snow reasonably well. If the snow has not melted naturally at the time of reading, the procedure is to melt the snow in the funnel and to measure the resultant liquid water in the usual way.

If amounts of snow are greater, with a risk of overcapping the orifice, the funnel might be heated to melt the snow as it falls; alternatively, the funnel can be removed so that the snow falls directly into an open container, the rim being retained to define the catch area. Without some means of melting the snow, however, even a gauge of this design will become full quickly. Some open gauges use antifreeze to melt the snow but they suffer from greater evaporative losses than those with funnels, although this can be prevented by floating a thin layer of light oil on the melted snow.

Measuring snow with recording raingauges   If falls do not overcap the funnel or bury the gauge, and natural melting is relied on, siphoning, weighing and tipping bucket gauges will all measure snow satisfactorily, provided that their mechanisms do not freeze up. The timings of the record will indicate melting time, not falling time, even though the total may be correct. If their funnels, internal mechanisms and outflow drains are heated, all will operate as if they were measuring rain. But heating a funnel to supply the necessary energy to melt the snow takes from 250 to 1000 watts.

Measuring snow automatically with the funnel removed   By removing the funnel, just as with a manual gauge it is possible to collect the snow directly in a large, open container, the build-up of snow being weighed mechanically or electronically. Simpler designs do not melt the snow, they simply hold it, and these are limited in capacity, unless natural melting occurs. More complex

Figure 8.11. Dry snow and ice crystals, 'spindrift', blow like sand, as here on Deception Island on the edge of Antarctica. Caught by raingauges, spindrift can be mistakenly interpreted as snowfall.

types of this kind use antifreeze to melt the snow as it falls into the bucket, with a film of oil to prevent evaporation. A further refinement is that of emptying the container and automatically replenishing the antifreeze, retaining the oil film by stopping siphoning in good time. One such design (intended primarily for rain measurement) omits the oil, measuring the losses as well as the gains in weight, thereby allowing for evaporation. But unless the snow is melted, the collector will soon become filled with snow and without natural melting will be limited in its capacity.

In all open (funnel-less) gauges there must be a danger that snow will get between the fixed outer rim and the internal container, possibly bridging it and then freezing; blowing snow finds its way through the smallest of holes and cracks.

Because of their ability to measure snow more easily than can a siphoning raingauge, weighing raingauges are probably more commonly used in countries that receive a significant proportion of their precipitation as snow.

Measuring snow with an optical precipitation gauge   Optical gauges also measure snow. Since they do not catch or melt the snow, many of the problems are avoided and it could be seen as a possible answer to the snowfall measurement problem. And so it is, partly. But while the water content of raindrops is precisely known, from their size, thereby allowing a moderately plausible estimate of rain intensity to be inferred from the scintillation signal, snowflakes vary widely in shape and density, making the conversion from light signal to intensity of snowfall much less certain than for rain.

Snow shields   Because of the large wind errors in measuring snow, attempts have been made to shield raingauges when they are used for snow, the Alter shield being an early example. The best measurement of snowfall using rain-gauges was deemed by the WMO to be that obtained by having bushes around the gauge cut to gauge height, or to site the gauge in a forest clearing. But bush screens are only practicable if the bushes do not become full of, or covered by, snow. Also, because suitable forest clearings or bushes may well not be available at many cold sites, and as they will all be different, they were not selected as the reference. Instead, and after due consultation and review, the WMO designated the *octagonal, vertical, double-fence shield* as the international reference (DFIR) against which all snow gauges should be compared; it is a secondary standard (Fig. 8.12). Artificial screens can also be moved as the snow accumulates, whereas plants cannot. A comparison of the DFIR was made with bush shields at Valdai in Russia, the bush exposure catching most snow, the DFIR 92% of this. All other gauges caught significantly less, for example an unshielded 8 inch manual gauge caught 57% of the bush gauge while an identical gauge with an Alter shield caught 75%. Methods have been developed to correct the DFIR shield to the bush readings (Yang *et al.* 1994) and these can then be compared with the catch of national gauges in an attempt to correct the national readings. This was done for 20 different gauges and shield combinations (Metcalfe & Goodison 1993). An example of the findings of the comparison is that the catch of Hellman-type gauges can be as little as 25% of the corrected DFIR gauge reading in winds above $5\,\mathrm{m\,s^{-1}}$ (Gunther 1993). With correction procedures for raingauges (see earlier) it is necessary to know wind speed and drop size; corrections for snow readings are likely to be much less reliable than for rain.

   Although correction procedures and field tests comparing national gauges with the DFIR are necessary, so as to be able to use the already existing data, the large size and expense of the double snow fence makes it impractical for general use – it is an equivalent, for example, to the pit gauge for rainfall. Since raingauges are, in so many ways, an unreliable means of measuring snow, alternatives have been sought.

Measurement of snow after it has fallen

Snow sections   After seven years of field tests 150 years ago, the only way that Huddleston could see of getting sensible readings was to take a 'cheese' of snow

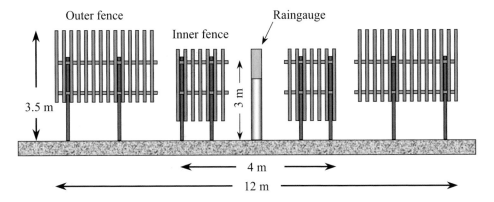

Figure 8.12. While wind effects can cause large errors in rainfall measurement, the problem is even greater with snow. To help mitigate these effects, the double-fence snow shield is the WMO-recommended international reference. A high outer octagonal slatted fence surrounds a similar but lower inner fence, the raingauge being at the centre, sometimes incorporating a further shield of the Alter or Nipher type (parts of the structure are omitted from the drawing for clarity).

off the ground where it seemed to be of average depth and to melt it. This still represents one of the best ways of measuring snowfall – but it needs an observer. After small falls of snow it is possible to collect a sample by inverting a raingauge funnel and pressing it through the snow until it meets the ground. The snow in the funnel is then melted and the water measured. If there is doubt about whether the chosen site is typical, several samples can be taken some distance apart, the measurements being recorded separately and the mean quoted.

If the snow depth is greater than about 15 cm and an area can be cleared next to that which is to be measured, it is then possible to insert a thin metal or wooden sheet horizontally into the snow pack at a suitable height from the top and to press the inverted gauge down until it meets the plate. The process is then repeated as many times as necessary to reach the ground.

Alternatively, a longer tube of known diameter and with a sharp edge can be pushed down into the snow and withdrawn with the snow core in it. The whole is then weighed and the known weight of the tube subtracted, the weight of the snow being converted into the water equivalent. This method does not work too well with wet snow because some water may be pressed out in the process. When new snow falls on old, a board, coloured white and with a slightly rough surface, is placed on the old snow just level with its surface and one of the above procedures carried out after the new snow has fallen. It is important not to trample the snow when collecting samples but to leave it undisturbed by following one path to the sample site every time. It is best to measure the collected snow soon after it has fallen to avoid evaporation.

Snow depth   It is relatively easy to measure the depth of snow manually, graduated poles fixed in the ground (or snowpack) simply being read at intervals. Automatic level sensors have also become available recently, sensing the top surface of the snow by downward-looking ultrasonic (echo) systems. They typically measure the depth to about ± 2 cm, although different manufacturers quote different values. Falling snow and spindrift can interfere with the readings of these echo-sounders, spurious signals being received as noise from the flakes. The air temperature also has to be measured to compensate for the variation in the speed of sound with temperature.

However, the snow's density is unknown, varying considerably from place to place, with the type of snow, and with time, and, as it is the water-equivalent of the snow that is usually required, without a knowledge of snow density this information cannot be deduced from the depth. Various rules of thumb are used regarding the density, such as the one-tenth rule, but snow density can vary from one tenth to one fiftieth, an average being about one thirteenth. Depth measurement alone is, therefore, only an approximate method. Taking test samples occasionally, to establish the density of the nearby snow, improves the estimate but is laborious and cannot be automated.

Snow weight   A way around not knowing the density of the snow is to measure the weight of the snowpack where it lies on the ground. Devices called *snow pillows* are used to do this. These are thin, flat containers made of flexible sheets of stainless steel, rubber or plastic, about a metre square and a centimetre or so thick, with a capacity of perhaps 50 litres of liquid. The liquid may be oil or a solution of antifreeze, the weight of snow being measured with a manually read manometer, electronic pressure sensor or float-type level recorder in a standpipe.

However, the snowpack changes with time and in particular can become solid and form ice bridges so that gradually its weight no longer rests fully on the pillow. A larger pillow, sometimes in the form of an array of four operated in parallel, can reduce but not eliminate this problem. If the snow does not stay for long and is not deep, bridging and solidification may not have time to occur.

Gamma ray attenuation   If a collated beam of gamma rays passes through the snowpack, it becomes attenuated to an extent depending on the water-equivalent of the snow. This is an effective, if expensive, technique but even though the gamma source is low level (typically 10 mCi of caesium 137), health concerns can cause problems. Consequently it is not in wide use. Natural background radiation is also sometimes used in a similar way.

Measurement of snow as it melts

Melting snow is of concern to hydrologists and to river managers, so information on melt rate is well worth having even if, due to evaporation and spindrift, it does not tell us how much fell in the first place, or when it fell. *Snowmelt lysimeters* collect and measure the meltwater flowing from the bottom of, or at some level within, a snowpack and take two forms – unenclosed and enclosed – the former being much the commoner (Fig. 8.13). In this design the collector has a shallow wall around it, while in the latter the wall completely isolates the snow column from its surroundings all the way up to the surface of the pack. In the latter design, the wall may be raised slowly as the snowpack increases in depth, although this is not of course possible at an unattended site. Either type may operate at atmospheric pressure or under negative pressure (as can soil lysimeters for evaporation measurement).

Meltwater percolating down through the ice pack, which is normally layered through alternate melting and freezing, with new snow being added periodically, follows anything but a direct downwards route, travelling laterally along the less permeable layers until it finds a way through (Wankiewicz 1979). The lysimeter intercepts this spatially variable flow but, as most lysimeters are small in area, what they collect may not be representative. The bigger the lysimeter, therefore, the better; the collecting area varies in practice, from less than 1 m^2 up to 100 m^2.

Meltwater flows mostly under the influence of gravity (Colbeck 1972, 1974), capillary effects being negligible although these may increase at discontinuities (Wankiewicz 1979). But a snowmelt lysimeter is just such a discontinuity, the layer at atmospheric pressure within the snowpack causing an *end effect*. This results in a saturated layer of up to 2 to 3 cm thick (Kattelmann 1984), where water is held in the pore spaces, between the snow grains, by capillary force; once formed, this saturated layer remains fairly constant. Above it is a pressure gradient zone, which decreases from atmospheric pressure at the lysimeter's supporting base to a negative pressure not far above. Percolating water may avoid this region and go round the lysimeter, but because the pressure gradient zone is only a few centimetres deep it can be contained within a lysimeter wall of from 12 to 15 cm, and so the problem can be avoided. However, the presence of this zone affects the timing of meltwater outflow from the lysimeter, since the saturated zone has first to form – a process that can take up to several days

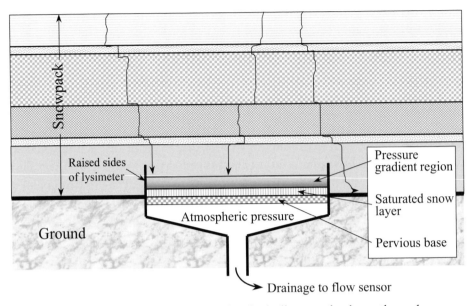

Figure 8.13. Meltwater percolating by indirect paths down through a snowpack, which comprises snow layers of varying texture and age. The volume of the meltwater can be measured by collecting it in a snowmelt lysimeter, the water being piped to a measuring device such as a tipping bucket. Provided that the sides of the lysimeter are higher than about 12 to 15 cm, the water is successfully collected, being unaffected by the pressure gradient introduced by the discontinuity that the presence of the collector generates. Because percolation can be spatially quite variable, the larger the area of the collector the better.

initially. Thereafter, the response to percolating water can be within hours or even minutes. The end-effect problems can be largely overcome by using tension lysimeters, in which a porous plate is held at a negative pressure approximating to that of the capillary pressure of the overlying snow. This much reduces the start-up and response times.

Enclosed lysimeters are useful for more precise work since they eliminate the lateral flow of meltwater at natural discontinuities within the snowpack. However, they are much less commonly used and their own, specialised problems will not be discussed here.

The meltwater collected by a lysimeter, piped away to a suitable site, can simply be read manually as it collects in a container, but more usually it will be measured automatically. The level of water in a container, perhaps sited slightly downhill if the installation is in mountains, can be measured with a float or pressure sensor, the water being siphoned off periodically. A weir tank with a water level recorder is an alternative if amounts of water are large, while weighing methods are also used, but a tipping bucket raingauge is the most common method. In all cases it is necessary to prevent freezing of the outflow pipes from the lysimeter, and of the measuring system, and this is helped by using large-diameter pipes and electrical heating if necessary.

Precipitation measurement using radar

Its place in environmental monitoring

This is a specialised technique in the area of *remote sensing* (Chapter 14) and, while beyond the main terms of reference of this book, which is concerned essentially with measurements made *in situ* on the ground by individual instruments, it is useful to describe how *weather radar* works, what its advantages and limitations are and how it relates to *in situ* raingauge measurements.

Radar was developed for the detection of aircraft, and the effects of precipitation on the received signal were originally seen as an inconvenient source of interference. However, even fifty years ago it was recognised that radar could be used to detect precipitation. Indeed Kurtyka (1953) remarked that 'In the last five years, the necessity of adequate precipitation instruments to calibrate radar for precipitation measurement has pointed to the primitiveness of the present-day rain gauge.' He went on to say that 'In all likelihood, the rain gauge of the future may be radar, for even in its present developmental stage, radar measures rainfall more accurately than a network of one raingauge per 200 square miles'. Well, radar has not replaced raingauges, and while it has advanced over the years it still benefits from, and relies on, *in situ* data from telemetering raingauges to calibrate the system, although this need may eventually be overcome.

Advantages of radar

One of the great advantages of radar is that it gives an areal estimate of precipitation rather than a single-point measurement, and the area covered is quite large, typically about $15\,000\,\mathrm{km}^2$ or more for each station. It also has the advantages of giving data in real time and of not needing anything to be installed in the area, or even access to it.

Principle of operation

A classic weather radar system, which had been very thoroughly evaluated by a consortium of interested groups (North West Water, the Meteorological Office, the Water Research Centre, the Department of the Environment and the Ministry of Agriculture, Fisheries and Food) and which was the first unmanned weather radar project in the UK, was installed on Hameldon Hill about 30 km north of Manchester during 1978. This brought within range the hills of North Wales, the southern Lake District and the Pennines and so gave the opportunity to evaluate a system in different terrains (Collier 1985). Much of the information below refers to this particular installation, but it is representative of the genre.

Weather radar uses a conical radar beam of 1 or 2 degrees, usually in the C-band (4 to 6 GHz), and measures the energy reflected and scattered back from the precipitation, converting this into an estimate of surface precipitation intensity. Estimates of the rainfall rate R are derived from the returned energy or *radar reflectivity* Z, using an empirical equation of the form $Z = aR^b$. Both a and b have many possible values, the most usual being $a = 200$, $b = 1.6$. b is usually left at about this figure while a may lie anywhere from 140 for drizzle, through 180 for widespread rain to 240 for showers. These are figures for the UK. In other situations they can be different, for example 500 for thunderstorms in an Alpine setting. These values are quoted to illustrate the magnitude of the variation of the reflected signal from different rain types and at different places; yet other values of a apply to snow.

Causes of uncertainty of measurements

The uncertainty in converting from reflectivity to precipitation intensity, and its variation from one place to another, is due to many factors, including the variation in drop-size spectrum and the presence of hail and snow. In addition, melting snow (snow turning into rain) increases the reflectivity, giving what is called the *bright band*, leading to overestimation of the rain reaching the surface. Conversely, drizzle tends to be underestimated, owing to the absence of large drops.

The second main cause of uncertainty is due to changes in precipitation intensity between the radar beam and the ground. A low beam elevation is necessary to detect rain as near to the ground as possible and to cover as long a range as possible, but hills can interfere with too low a beam. In the case of the Hameldon Hill installation, four elevations of beam were used: 0.5, 1.5, 2.5 and 4 degrees. The two lower angles sense surface precipitation, the 0.5 degree beam having a range of 24 km and the 1.5 degree beam being used for distances beyond. The two higher elevations are for the detection of the bright band (melting snow) and for other studies.

The antenna dish rotates at a speed of about one revolution per minute, scanning each elevation in turn and so giving around one scan of the surface rainfall every three minutes. However, the need to direct the beam slightly upwards can result in the beam missing events below it, such as shallow precipitation or the orographic enhancement of rainfall due to the presence of hills. Furthermore, even with an appropriate upward elevation of the beam, reflections from the ground or sea can still occur, particularly in the presence of a strong hydrolapse (dry air over moist air), which causes the beam to bend downwards (Fig. 8.14). Since the beam scans round, it does not take continuous readings at any one point but samples each direction and distance briefly as it passes. The measurements are thus spot instantaneous readings of intensity taken about once every three minutes over any one particular location. This snapshot view inevitably adds perhaps 10% to the uncertainty of conversion from spot observations to mean values, which arises from the temporal and spatial variability.

Corrections can be made for permanent ground obstructions and for the reduced beam depth caused by intervening hills (although what is happening beyond and below the hills cannot be seen). Allowance is also made for a reduction in outgoing and returned signals due to their absorption as they pass through heavy rain. And, of course, allowance has to be made for the reduction in signal due to the inverse square law. At longer ranges, allowance also has to be made for earth curvature and the fact that a significant part of the beam may, therefore, be above the rain.

An improvement in the conversion from reflectivity to intensity can be obtained by adjusting the value of a, in the empirical formula, through the use of *in situ* raingauges telemetering an input of 'ground-truth'. Over the last ten years, however, there has been extensive development of adjustment techniques and while there are still problems, the reliability of estimates has much increased.

Accuracy and corrections

The accuracy achieved by the radar system operated on Hameldon Hill for rainfall measurements was tested by comparing the hourly values of the radar estimate, for the areas containing a raingauge, with the hourly raingauge readings (Collier 1986). The radar estimate was the average of 12 instantaneous readings of intensity integrated to give hourly totals. Individual differences between the radar and raingauge measurements varied very considerably from the mean. When observations influenced by the bright band were excluded, the average difference, under conditions of frontal rain, was 60%, and 37% for convective rain. If bright band episodes were not excluded, the difference could be 100%.

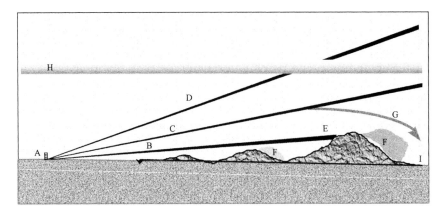

Figure 8.14. In this schematic of a weather radar system, A is the radar transmitter, which radiates a beam of width 1 to 2 degrees. Beams of a low angle of elevation (0.5 degrees, B, and 1.5 degrees, C) sense precipitation close to the ground (B up to, and C beyond, 24 km), although they can miss orographic rainfall caused by hills below the beam (F); as distance increases, the fall-off of the ground due to the curvature of the earth (I) increases the height below which events cannot be measured. The beams can also be obstructed by hills, which cause shadows beyond them (E). In the presence of a hydrolapse, the radar beam can be bent towards the ground, giving false reflections (G). A higher beam (D) is used to detect the bright band (H) caused by melting snow, which gives anomalously high reflected signals.

To help reduce the error, ground-truth from telemetering raingauges can be used. However, if only a few gauges are used – to give an average correction for the whole area – it is unlikely to help, because of spatial and temporal variations. An alternative is to use a large number of gauges to sense more widely the spatial variations in the difference between radar and raingauge readings, but this requires too many gauges.

A less direct method is based on the fact that automatic detection of the *type* of rainfall allows the most appropriate calibration procedures to be used. Detection of rainfall type can be achieved if the temporal variability of the difference between readings is measured at a small number of raingauge sites, for it has been found that this variability can identify the type of rainfall well enough, and so can be used to choose the best correction factor. This approach lessens the difference between radar estimates and raingauge measurements from 60% without calibration to 45% for frontal rain, and from 37% to 21% for convective rainfall. Over the last decade much work has been done to improve the conversion from radar signal to estimated rainfall. For those needing to know more, there is a detailed review by Collier (1996).

More sophisticated techniques could improve the performance of radar estimates, such as the use of beams polarised in the vertical and horizontal directions, whereby it is possible to detect the raindrop-size distribution more precisely. The reason is that raindrops are nearly oblate spheroids, with their axes of symmetry aligned close to the vertical. The degree of oblateness can be detected by the polarised beams and this relates to drop size. However, there are problems with this method, not the least being an increase in the complexity of data processing and of hardware, and thus of cost. So far this type of radar has not been introduced operationally. There are a number of other new techniques using polarisation-diversity radar, the most promising seeming to be the use of *differential attenuation* and *differential phase* (Collier 1997).

Another possible avenue of enhancement of radar rainfall measurement lies in a project known as FRONTIERS, which combines radar precipitation measurements with remotely sensed data from satellites (Chapter 14).

Snowfall measurement by radar

Since snowfall is difficult to measure by other means it would be useful if radar could do better. The same empirical equation is used for snow as for rain, and as in the case of rainfall there is considerable variability in the values of *a* and *b*, *a* lying somewhere between 2100 for wet snow and 540 for dry. However, the same problem exists for radar as for optical raingauges – the size and density, and thus the water equivalent, of snowflakes vary so much that there is a considerably larger margin of uncertainty than for rain. Just as spindrift is a problem when trying to measure snowfall with collectors, it presents a problem when using radar since the subsequent movement of fallen snow by wind takes place below the beam and so goes undetected.

There is also the additional problem that because it is extremely difficult to measure snowfall by any conventional means there are no reliable measurements against which to compare the radar's snow readings, either for checking performance during development or as real-time ground-truth in the operational state. It is possible, however, to use measurements of snow made after it has fallen to assess the radar's performance in retrospect. Because snowflakes are not spheroid-shaped, polarisation techniques are inappropriate. Snowfall measurement is just a very intractable problem.

Displaying the rainfall measurements

Although radar positions are given in polar co-ordinates, they can be converted to Cartesian co-ordinates for display on a monitor screen in the form of a square grid. A typical size of grid will vary from 2 km for ranges up to 75 km to 5 km for ranges from 75 to 200 km. But modern radars allow the user to retain polar grid displays if preferred. Levels of precipitation intensity are displayed using different colours. In addition, rainfall totals over areas of interest, such as river catchments, can also be calculated over a period of, say, 15 minutes.

Cost

The capital cost of a weather radar system is high – hundreds of thousands of dollars – and being complex it needs high-level technical expertise to maintain it. Weather radar is thus best suited to situations where measurements are needed from a large area on a permanent basis in real time, such as for a national weather service or a large river authority. Radar is not suited to measuring precipitation over smaller areas or for short durations, although this kind of information may be purchasable by smaller organisations from those operating large radar networks.

References

Alter, J. C. (1937) Shielded storage precipitation gauges. *Mon. Weather Review,* **65,** 262–5.

Calder, I. R. & Kidd, C. H. R. (1978) A note on the dynamic calibration of tipping bucket raingauges. *J. Hydrol.,* **39,** 383–6.

Colbeck, S. C. (1972) A theory of water percolation in snow. *J. Glaciology,* **11,** 369–85.

Colbeck, S. C. (1974) The capillary effects on water percolation in homogeneous snow. *J. Glaciology,* **13,** 85–97.

Collier, C. G. (1985) Accuracy of real-time estimates made using radar. In *Proc. Symp. Weather Radar and Flood Warning,* University of Lancaster 1985.

Collier, C. G. (1986) Accuracy of rainfall estimates by radar, part 1: calibration by telemetering raingauges. *J. Hydrol.,* **83,** 207–23.

Collier, C. G. (1996) *Applications of Weather Radar Systems. A Guide to Uses of Radar Data in Meteorology and Hydrology, second edition.* John Wiley/Praxis, Chichester, 390p.

Collier, C. G. (1997) Private communication.

Danby, H. (1933) *Translation of 'The Mishnah'.* Oxford University Press.

Dobson, D. (1777) Observations on the annual evaporation at Liverpool in Lancashire; and on evaporation considered as a test of the dryness of the atmosphere. *Phil. Trans. Roy. Soc. London,* **67,** 244–59.

Essery, C. I. & Wilcock, D. N. (1991) The variation in rainfall catch from standard UK Meteorological Office raingauges: a twelve year case study. *J. Hydrol. Sci.,* **36,** 23–34.

Folland, C. K. (1988) Numerical models of the raingauge exposure problem, field experiments and an improved collector design. *Quart. J. Roy. Met. Soc.,* **114,** 1485–516.

Genrich, V. (1989) Introducing 'electronic impact sizing' as a new, cost-effective technique for the on-line evaluation of precipitation. In *Proc. WMO/IAHS/ETH Int. Workshop on Precipitation Measurement,* St Moritz, 211–16.

Grew, N. (1681) In *Musaeum Regalis Societatis,* London 357–8.

Gunther, Th. (1993) German participation in the WMO solid precipitation intercomparison: final results. In *Vol. 1, Proc. Symp. Precipitation and Evaporation, Bratislava 1993,* pp. 93–102.

Hanna, E. (1995) How effective are tipping-bucket raingauges? A review. *Weather,* **50,** 336–42.

Heberden, W. (1769) Of the quantities of rain which appear to fall at different heights over the same spot of ground. *Phil. Trans. Roy. Soc.,* **59,** 359.

Hewston, G. M. & Sweet, S. H. (1989) Trials use of weighing tipping-bucket raingauge. *Met. Mag.,* **118,** 132–4.

Huddleston, F. (1933) A summary of seven years' experiments with raingauge shields in exposed positions, 1926–1932 at Hutton John, Penrith. *British Rainfall,* **73,** 274–93.

Hughes, C., Strangeways, I. C. & Roberts, A. M. (1993) Field evaluation of two aerodynamic raingauges. *Weather,* **48,** 66–71.

Jevons, W. S. (1861) On the deficiency of rain in an elevated raingauge as caused by wind. *London, Edinburgh and Dublin Phil. Mag.* **22,** 421–33.

Jones, S. B. (1983) The estimation of catchments average point rainfalls. Institute of Hydrology Report No. 87.

Kattelmann, S. C. (1984) Snowmelt lysimeters: design and use. In *Proc. Western Snow Conf.,* pp. 68–76.

Koschmeider, H. (1934) Methods and results of definite rain measurements III. Danzig Report (1). *Mon. Weath. Rev.,* **62,** 5–7.

Kurtyka, J. C. (1953) Precipitation measurement study. Report of investigation No. 20, State Water Survey, Illinois, p. 178.

Metcalfe, J. R. & Goodison, B. E. (1993) Correction of Canadian winter precipitation data. In *Proc. 8th Symp. Met. Obs. and Insts. 1993, Anaheim, CA,* pp. 338–43.

Nipher, F. E. (1878) On the determination of the true rainfall in elevated gauges. *Am. Assoc. Adv. Sci.,* **27,** 103–8.

Norbury, J. R. & White, W. J. (1971) A rapid-response raingauge. *J. Phys. E, Scientific Instruments,* **4,** 601–2.

Reynolds, G. (1965) A history of raingauges. *Weather,* **20,** 106–14.

Robinson, A. C. & Rodda, J. C. (1969) Rain, wind and the aerodynamic characteristics of raingauges. *Met. Mag.,* **98,** 113–20.

Rodda, J. C. (1962) An objective method for the assessment of areal rainfall amounts. *Weather,* **17,** 54–9.

Rodda, J. C. (1967a) The rainfall measurement problem. In *Proc. Bern Assembly IASH,* pp. 215–31.

Rodda, J. C. (1967b) The systematic error in rainfall measurement. *J. Inst. Water Eng.,* **21,** 173–9.

Sevruk, B. & Hamon, W. R. (1984) International comparisons of national precipitation gauges with a reference pit gauge. WMO Instruments and Observing Methods, Report No 17. WMO/TD, No. 38.

Sevruk, B. & Klem, S. (1989) Types of standard precipitation gauges. In *WMO/IAHS/ETH Int. Workshop on Precipitation measurement, St Moritz.*

Sevruk, B. (1996) Adjustment of tipping-bucket precipitation gauge measurements. *Atmos. Res.,* **42**, 237–46.

Sevruk, B., Hertig, J. A. & Spiess, R. (1991) The effect of a precipitation gauge orifice rim on the wind field deformation as investigated in a wind tunnel. *Atmos. Environ.,* **25**, A(7), 1173–9.

Shamasastry, R. (1915) Translation of 'Arthasastra' by Kautilya. Gov. Oriental Library Series, Bibliotheca Sanskrita, No. 37, part 2, Bangalore.

Stevenson, T. (1842) On the defects of raingauges with descriptions of an improved form. *Edinburgh New Phil. J.,* **33**, 12–21.

Strangeways, I. C. (1984) Low cost hydrological data collection. In *Proc. IAHS Symp. on Challenges in African Hydrology and Water Resources, Harare,* Publ. No. 144, 229–33.

Strangeways, I. C. (1996) Back to basics: The 'met. enclosure': Part 2(b) – Raingauges, their errors. *Weather,* **51**, 298–303.

Symons, G. J. (1864) Rain gauges and hints on observing them. *British Rainfall,* 8–13.

Symons, G. J. (1866) On the rainfall of the British Isles. In *Report on 35th Meeting of the Brit. Ass. for the Adv. of Sci., Birmingham,* 1865, pp. 192–242.

Thiessen, A. H. (1911) Precipitation for large areas. *Mon. Weather Review,* **39**, 1082–4.

Townley, R. (1694) Observations on the quantity of rain falling monthly for several years successively. A Letter from Richard Townley. *Phil. Trans. Roy. Soc. London,* **18**, 52.

Wankiewicz, A. (1979) A review of water movement in snow. In *Proc. Modelling of Snow Cover Runoff, US Army Corps of Eng. Cold Regions Research and Eng. Lab., Hanover, US,* pp. 222–52.

Weisinger, T. (1996) Aerodynamically superior precipitation gauges. Report to CEN/TC 318 Working Group 5 (British Standards Institute, London).

Whittaker, A. E. (1991) Precipitation rate measurement. *Weather,* **46**, 321–4.

Yang, D., Sevruk, B., Eloma, E., Golubev, B., Goodison, B. & Gunther, Th. (1994) Wind-induced error in snow measurement: WMO intercomparison results. In *Proc. 23rd Int. Conf. on Alpine Meteorology, German Weather Service,* **30**, pp. 61–4.

9

Soil moisture and groundwater

Subsurface water processes

Soil moisture

Soil moisture (or soil water) refers to the water that occupies the spaces between soil particles. It is at its maximum when the soil is saturated, that is when all the air between the particles is replaced by water but, if the soil can drain, the spaces will normally also contain air, the water then forming a thin film on and between the soil particles, held by capillary attraction. As the soil dries out this film becomes thinner and progressively less easy for plant roots to extract. The water is free to move through the soil, up or down, by gravity and by capillary attraction; it is taken up by plant roots, evaporates at the surface or recharges the groundwater.

However, there is also water present in soil which is not free to move or to be taken up by plants but which may nevertheless be detected during measurement – but not differentiated from the free water, depending on the measurement technique used. This is the *water of crystallisation* or *water of hydration* – water that is chemically bound to minerals within the soil such as gypsum ($CaSO_4\ 2H_2O$); water may also be bonded to organic material to varying degrees of strength. In making measurements of soil water content, soils of great variety are encountered and any instrument or measurement method must be able to handle them all, from pure peat or sand to silt and clay or a mix of them, all varying in pore size or having a variety of pore sizes combined, and all varying in the extent of chemically bonded water. Anomalies, such as stones dispersed at random in the soil profile, can also disrupt measurements since stones do not usually hold any water and if not detected can give rise to large errors, again partly depending on the method of measurement. The terms 'soil moisture' and 'soil water' are generally interchangeable although the latter could be interpreted as including the chemically bound water, which, as explained above, is not in the form of moisture.

Soil tension

The spaces between the soil particles contain air, water and roots. When the soil pores contain only water then the water can move freely, the largest force acting on it being that of gravity which causes it to move downwards provided there is drainage. It then drains away until an equilibrium is reached known as the *field capacity*, at which point any further water added cannot be retained and will drain away. As the soil dries out (for whatever reason), the water in the pores decreases in volume and retreats to increasingly small spaces between the particles, becoming held more and more tightly by capillary attraction. Under anything less than field capacity there is said to be *soil moisture deficit*. As drying occurs, the suction that plant roots must apply to extract water increases. This suction or *soil tension* (or *soil water potential* or *soil matrix potential*) is measured in the same units as barometric pressure. At saturation (no air spaces), the tension is zero while at field capacity it is around -0.3 bar (-300 hPa). As drying continues, the remaining water is eventually held so tightly that it is not possible for plants to apply enough tension to extract it, this state being known as the *permanent wilting point*, occurring at about -15 bar. The exact value differs between plant types, but at this point they die. However, most of the water available to plants is held at pressures of less than -1 bar, although negative pressures can rise as high as -60 bar. In soil physics, a non-SI unit of pressure, *centimetres of water*, is often used as it is more directly meaningful (1 bar $= 1017$ cm of water; see Chapter 7).

Knowing the strength of soil tension is of value to agriculture, since it acts as a warning of when it is necessary to start irrigation as well as when it is wasteful to continue. But the significance of soil tension is also due to the control it exerts on the movement of water within a soil profile. If the combined upward tension exceeds the pull of gravity, it is possible for water to move upwards and be evaporated or transpired. If gravity and downward tension combined are the stronger force, water moves downwards, eventually joining the saturated zone of groundwater at the water table. But temperature gradients within the soil also cause water to move (Gilman 1977), liquid water moving from colder to warmer zones, vapour from warm to cold (by evaporation and condensation). This was understood many years ago (Patten 1909).

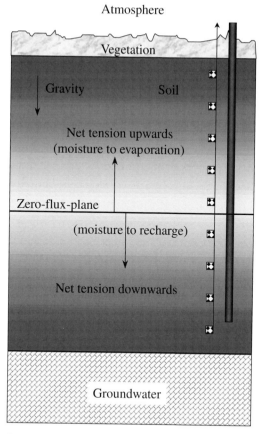

Figure 9.1. The zero-flux plane. Depending on conditions, there can be a depth in the soil through which the flux of soil moisture is zero, moisture above moving upwards to evaporation and moisture below moving downwards to groundwater recharge. The presence and position of the plane depends on the net force on the soil water of gravity and soil tension. The depth of the plane can be established by installing tensiometers at several depths (the vertical array of squares), and the amount of soil moisture at these depths can be measured by neutron probe (the access tube is shown on the right of the array of tensiometers).

The zero-flux plane

The point in a soil profile at which the water above flows up and the water below flows down is known as the *zero-flux plane* (ZFP) – the plane through which no soil moisture is passing at the time of the measurement (Fig. 9.1). The position of the plane moves up and down depending on conditions, or it may not exist at all (for example after heavy rainfall in summer, or throughout the winter – all movement then being downwards). Knowing the plane's position is useful since it divides the soil moisture into that which is being evaporated or transpired and that going to recharge the groundwater and eventually to contribute to river flow. The position of the ZFP can be measured by sensing the relative values of tension in the soil at different depths. Further, if the ZFP is measured in addition to the moisture content in the same profile, the amounts of water going up and down can be quantified.

Groundwater

The part of the soil moisture which does not rise upwards and evaporate instead percolates downwards and joins the groundwater – a region in which the soil or permeable rock is saturated with water without airspaces. From here, in due course, the water finds its way into a drainage ditch, stream, river or lake and eventually to the sea. Measuring the depth of the water table, changes in its level and the quality of the groundwater is of importance not just to researchers investigating hydrological processes but to those concerned with the practical matters of water resources, civil engineering, agriculture and pollution.

Units and terms

The simplest method of measuring soil moisture is to take a sample of soil, weigh it, dry it and weigh it again. This basic method is described more fully later, but is introduced here because it allows the units of measurement to be understood more graphically.

There are two ways of expressing the amount of moisture present in a sample of soil, one giving the value in terms of the weight of the moisture, the other in terms of its volume. Suppose a sample is taken roughly with a spade – with no knowledge of the volume it occupied *in situ* – and that this sample is weighed before it has lost any moisture. It is then dried in an oven and weighed again. The loss of weight (loss of water) is divided by the dry weight of the soil and the result is known as the *moisture dry-weight fraction* (MDWF). By multiplying the fraction by 100, the result can be given as a percentage.

But it is generally more meaningful to express the amount of moisture in terms of the volume, not the weight. The result is then expressed as the *moisture volume fraction* (MVF) or *volumetric water content* (VWC), the volume of the water divided by the volume of the wet soil. Again if multiplied by 100 the result is expressed as a percentage – the *moisture volume percentage* (MVP). In these terms, 100% indicates pure water with no soil, while 0% is pure soil containing no water. In the same terms, when all air spaces are occupied by water the MVP is about 35% to 40% for sand, through 65% for clay and as high as 98% for peat. The reason the MVP is often preferable to the MDWF is that it is then in units compatible with related variables such as rainfall and evaporation – that is, they are all expressed as a thickness (millimetres) of water gained or lost over a period. This helps in calculating a water balance.

As an example, if measurements have shown that most of the change in soil moisture occurs within the top metre of a particular soil, and by measurement it is found that the average MVP has changed by $+5\%$ over a month, then the change can be expressed as 5% of 1000 mm (1 m) in a month, or $+50$ mm of water; that is, the soil now holds 50 mm more water than it did. Perhaps it has retained 50 mm of the rainfall, or some moisture has moved upwards from below under the influence of tension. This can be compared with the rainfall, of say 200 mm, and with the measured evaporation of, say, 30 mm. From this, the amount left that will appear as runoff can be calculated:

$$\text{runoff} = 200 - 30 - 50 = 120 \, \text{mm}$$

If the area of the catchment is $1 \, \text{km}^2$ then the volume available as runoff is $1000 \times 0.12 \, \text{m}^3 = 120 \, \text{m}^3$. This amount will not run off into the river at a constant rate over the period of, say, one month, but will follow a curve known as a hydrograph (a plot of runoff against time), the shape of which will depend on how the 200 mm of rain fell (the time and intensity) and on the physical characteristics of the ground. The point to note is the advantage of expressing soil moisture as a fraction of the volume of soil, not as its weight.

Measuring soil moisture

The gravimetric method

The gravimetric method was the only way of measuring soil moisture until the development of the electronic methods to be described later; it is now mainly used to calibrate the electronic methods, although it is still in routine use where costs must be kept to a minimum, or where only a few measurements for a short time are needed and the high cost of electronic instruments would be unjustified.

The basic procedure   A soil sample is taken, stored so that water cannot evaporate from it, returned to a laboratory, weighed and dried in a *ventilated* oven at 105°C until its weight is constant. It is then cooled and reweighed (hence the more correct name is the *thermogravimetric* method); it can take from 0.5 to 2 days to complete the drying process. This gives a measure of the weight of (free) water which the sample contained in the field. Because the density of water in c.g.s. units is as good as unity for all practical purposes, knowing the weight of water driven off by heating tells us the volume; for example, if the weight loss is 20 grams, the volume of water loss is 20 ml.

However, measuring the volume of the soil is more difficult. The usual method is to collect a sample with a *soil corer* – a sharp-edged cutting cylinder, pushed into the soil to a precisely known depth and then withdrawn with the soil inside it. But if the sample does not slide easily into the tube, soil can be pushed sideways and lost as the sample becomes compacted. Stones can prevent a sample being taken or they can deflect the corer. If the corer is moved from side to side to get it to go down against resistance, spaces can be introduced into the sample and small stones in the sample can introduce a large random error. It is also difficult to insert the tool to exactly the correct depth. When the corer is removed, a clean break (level with the end of the corer) rarely occurs. While some of these difficulties can be prevented by careful work and repeated attempts, even then the error in the volume, and so in the MVP, can be $\pm 10\%$.

A more precise method   If an error of this magnitude is unacceptable, more complex procedures can be adopted; one such involves the careful digging of a small hole of known dimensions (say 10 cm in diameter by 15 cm deep), by hand using a trowel, the soil being stored in a plastic bag for later weighing, drying and reweighing. The volume of the hole is then measured by lining it with a thin plastic bag into which water is added until level with the top of the hole, the volume of water poured into the bag being equal to the volume of the sample of soil (Bell 1996). This is very time consuming, but it has the added advantage that in addition to measuring the MVP it also allows the *dry bulk density* σ_d of the soil to be established at each depth at the site:

$$\sigma_d = \text{mass of the dry soil in grams} \div \text{volume of the dry soil in ml}$$

This needs to be done only once because σ_d is fairly constant for most soils, clays excepted; it ranges from just below 1.0 for topsoil to 1.75 for silts, sands and gravels. The advantage of knowing σ_d is that it can then be used in the third and most practical way of using the gravimetric method, which we now discuss.

The screw-auger method   Soil samples are collected at the various depths required (at which σ_d has already been established) using a screw auger. The MDWF (moisture dry-weight fraction) is then determined. In this method, the volume of the sample is not needed, for if σ_d is known then the MDWF can be converted to the MVF, since MVF = MDWF $\times \sigma_d$. For example, if the MDWF is 0.30 and σ_d is 1.5 then the MVF $= 0.30 \times 1.5 = 0.45$ (or 45% MVP).

Auger holes are easy to make down to over two metres in depth, and need only be 2.5 cm in diameter. Using this technique errors are most likely to occur in the course of establishing σ_d, so care is needed in that crucial initial measurement. Errors may also occur if σ_d varies much around the point at which measurements will be made, so several initial measurements of σ_d are a wise precaution even if time consuming and tedious. Errors in weighing will be very small in comparison.

The advantages of the gravimetric method are the cheapness of the equipment and the simplicity of the field work (after the initial careful measurement of σ_d), so that unskilled workers can be used. It is also an easy-to-understand method and there is nothing to go wrong with the equipment.

Limitations and sources of error The method assumes that the properties of the soil are fairly consistent within the area around the sampling point, but as with other variables there is always some spatial variation. Because the method is destructive, a different sample of soil is measured each time and so there will be some error due to this variation – it could be that the reason for a difference between measurements from one observation and the next is that the soil properties are different, not that there are moisture changes. Repeated measurements over time are normally what are required, because it is the changes in moisture content that are usually significant, for example in an estimate of catchment water balance or a measurement of the use of water by a particular crop. Some of the new electronic, non-destructive methods have the advantage of measuring the same volume of soil each time, and so they overcome the spatial variation problem – at least as far as comparing changes at the one point is concerned. (They do not overcome the problem that the sampling point may not be representative of the area. That is another problem, akin to that encountered with all variables when selecting representative sites, but just plain common sense goes a long way in this.)

To quantify the problem that spatial variation causes uncertainty about the interpretation of differences between consecutive readings at the site, it is possible to do a preliminary evaluation of the extent of this variability. About 10 sampling points are selected and samples taken at each depth at which measurements will routinely be made. Using the dry bulk density method, the MVP is calculated for each point and for each depth and the standard deviation determined for the readings at each depth. These can then be used either to estimate the accuracy that one single sample will give or the number of samples that needs to be taken to achieve the accuracy required (Bell 1996).

For better results, however, there are more complex (and more expensive) methods of measuring soil moisture, as follows.

The neutron probe

The neutron probe is arguably the best method available for the measurement of soil moisture, although two others, described later, the *capacitance probe* and the *time domain reflectivity* method (TDR), are also now available with their own particular advantages.

Principle of the neutron probe   A source of radioactive material, generally americium 241–berylium, with an intensity of between 30 and 100 millicuries (1 millicurie (mCi) = 0.037 GBq or gigabecquerels) and contained within a tube, emits fast neutrons into the soil (Fig. 9.2); americium–beryllium is used because it emits little gamma radiation and so is safer. As the fast neutrons move through the soil they progressively lose energy and become slowed and scattered as they collide with hydrogen nuclei. Most of the hydrogen atoms in the soil are those contained in water molecules. After many collisions and direction changes, some of the fast neutrons find their way back to the tube containing the radioactive source, as slow or 'thermal' neutrons (Fig. 9.2). The soil itself and some elements within it, such as cadmium, have a similar, though slight, effect. Also in the sensor tube is a slow-neutron detector – usually a boron trifluoride sensor, owing to its cheapness, stability and reliability –, which produces a pulse output for each slow-neutron that enters it, the number of such encounters being proportional to the number of water molecules in the soil.

The rate at which thermal neutrons are returning is measured and displayed by a *rate-scaler* over a period, typically of 16 or 60 seconds, the rate being displayed as counts per second. For more accurate readings the counting time can be extended to minutes. (Because the radioactive source emits neutrons randomly, the count rate will vary from one measurement to the next, even if the water content did not change, but by counting over a longer period this effect is minimised.) The half-life of the source is 450 years, so its gradual decline in intensity is not a factor. The count rate is displayed on a LCD screen, but can also be stored in the memory within a *recording rate-scaler* for later downloading, the site location of the reading also being stored in the memory.

For health safety, the tube containing the radioactive source is housed in a protective spherical shield made of a moderating material such as polypropylene. This reduces the radiation level outside to a value that is well within safety limits (less than 0.5 mrem h^{-1}), allowing the instrument to be carried by an operator all day without danger. Since the instrument often has to be borne

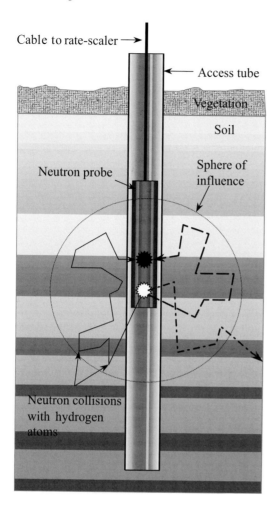

Figure 9.2. A neutron probe contains a source of fast neutrons (white sunburst), which are emitted into the surrounding soil and collide with the hydrogen atoms of soil moisture; some return, much slowed, to a slow-neutron detector in the probe (black sunburst). The resultant pulses are counted by a rate-scaler (above ground). Depending on the amount of soil moisture present, the sphere of influence varies from a radius of 15 to 30 cm. The solid, broken and broken-and-dotted lines show the paths of three neutrons.

across rough country, its weight needs to be minimised and a typical complete probe will weigh around 10 kg (Fig. 9.3).

Operation in the field   The probe is placed on an *access tube*, permanently installed in the ground to a depth appropriate to needs, which might be from near-surface to several metres (Bell 1976). The tube is usually of aluminium

Figure 9.3. In the foreground, a neutron probe is seated on an access tube, the probe having been lowered into it; a rate-scaler (folded open, to the right) indicates the rate of return of slow neutrons from the moist ground. Beyond, against the wall, is a tensiometer manometer board; see Fig. 9.11(*b*).

alloy, which is fairly transparent to slow neutrons (stainless steel absorbs them more and so reduces the count rate, as does brass although less so). A cap at the base prevents water or soil entering from below and aids installation, while a removable rubber bung at the top protects the probe from water, debris ingress and insects. The tubes are installed by first auguring out the soil to a diameter that allows the tube to fit as tightly as possible, in order to prevent water running down the outside of it, which would change the soil moisture below. As with all soil observations, it is important not to tread the ground down, which would compact it around the access tubes; wooden walkways are used to minimise the problem.

The probe is lowered from its shield into the access tube on a cable to each depth at which a reading is to be made, a counter indicating the depth that the detector has reached. At each depth, the rate-scaler is set counting and at the end of the period it displays the count rate, this being written down by the operator or automatically stored in the memory.

Calibration   The MVF, MVP or VWC are defined, for the purpose of calibrating a neutron probe, as the volume of water that would be removed from the soil by heating it to 105 °C – as in the gravimetric method. But, as noted earlier, not all the water is driven off by heating, some remaining as water of crystallisation and bound to organic material. This water affects the count rate as if it was free water but, as the amount is fairly constant for a given sample of soil, changes in count rate may be attributed to changes in water content. It is largely the changes, rather than an absolute measure of water content, that matter in practice.

Calibration can be done theoretically, or in a soil sample set up in the laboratory or in the field. Theoretical calibrations are based on a precise chemical analysis of the soil for such materials as cadmium, boron and chlorine that also scatter neutrons, but it is not much used since the analysis is expensive. Laboratory calibrations require a soil sample, weighing several tons, to be carefully dug from the field and repacked into a drum at least 1.5 m diameter and 1.5 m deep with an access tube down the centre. But it is only practicable if the soil is homogeneous and can be repacked to something resembling field conditions. In practice this is only useful for gravel, sand or silts.

The simplest method is field calibration, although it is slow and laborious. Near to the permanent access tube, a temporary tube is installed and, at half a dozen points around it, gravimetric measurements are made at each depth at which the probe will be used. This number of points is necessary to obtain a linear regression, because of the variability of the soil. Probe readings are taken at the same time at each depth. Because the readings must cover the full moisture range of the soil, readings need to be taken periodically throughout a full season and so it can take up to a year to obtain a full calibration.

The gravimetric soil moisture measurements are then plotted against the ratio of the probe count rate in the soil, R, to the count rate in a standard moderator, usually water, R_w. The latter is used instead of simply R in order to compensate for variations, from probe to probe, in the sensitivity of the detectors, the intensity of the neutron source and other characteristics. Figure 9.4 shows the type of calibration obtained. The lines are almost linear and this allows changes in MVP to be estimated reasonably accurately by the probe. The lines are also nearly parallel, and so one composite representative graph can be used for all soils.

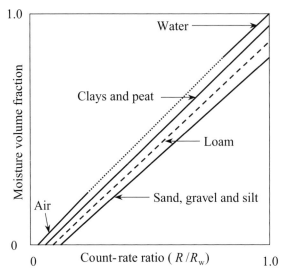

Figure 9.4. It is better to calibrate a neutron probe not in terms of the count rate in soil, R, but as the ratio of R to the count rate in water, R_w. Calibrations will vary according to the soil's chemistry and density, roughly as illustrated here. The lines do not go through zero because there will always be a small residual count.

The value of R_w is obtained by constructing a cylindrical container at least 60 cm deep by 50 cm diameter, with an access tube into which the probes can be lowered, to the centre. The container is filled with water. To minimise the random error a long count is made (perhaps 8 minutes). As a further check, a quick test is made at each field site before taking the actual soil readings, by taking a 16 second count with the probe in its shield (which acts as a moderator in a similar way to water – but less consistently).

Volume of soil measured by a neutron probe   The radius of the sphere over which the neutrons have influence depends not upon the strength of the radioactive source but upon its energy, so any probe using americium–beryllium will cover the same range. Tests have shown that, using this type of source, the effective radius of the probe is about 15 cm in wet soil and perhaps 30 cm in dry soil (Fig. 9.2). So the measured value of MVP is the average over a sphere with these diameters. This also means that greater resolution cannot be obtained by taking readings at levels closer together than about 15 cm, since the sampling volume then overlaps and some of the same soil is included in two adjacent readings. It also means that it is difficult to make reliable measurements in the top 20 cm of soil, as part of the neutron cloud is then in the air above ground. Ways around this include the use of 15 cm-deep *surface exten-*

sion trays – circular, open-bottomed trays, which are filled with the same soil and vegetation as the surroundings and which can be lifted from an open-bottomed shallow storage pit and placed around the access tube during readings.

Access-tube networks   Apart from situations where soil moisture is to be measured at one localised point for a specific reason, the aim is usually to obtain an areal estimate of soil moisture – just as in the case of rainfall – and the methods used in deciding the siting of access tubes can be the same as those used for selecting raingauge sites (Chapter 8). One of these techniques is to instal as many access tubes as can be conveniently visited in, say, a day; readings are taken quickly (in 16 seconds), high precision at a few points being less useful than lower precision at many, because of the great spatial variability of soil moisture. After a year or more, *index sites* may be identified that correlate closely with the network mean, and if this is the case many of the access tubes can then be discontinued and just the key sites used (as with raingauges). At these key sites, measurements can then be made more often and with longer counting periods (1 minute instead of 16 seconds, for example); see Bell (1987).

Problems   The neutron probe is a manual instrument carried to a site. The method could not be automated without developing a technique to move the probe up and down the access tube, and although this was considered in the early days of probe development, it would not be economic to commit a probe to every access tube in a network, made all the more expensive by the inclusion of automation. It might have been an option at a few selected sites, but safety considerations – with regard to leaving radioactive sources unattended at field sites – were a further deterrent. Safety considerations also make the probe difficult to transport, with warnings required on cars and problems when the probe has to be transported by air. There is also always the danger that in a technologically backward country a probe could be abandoned and later dismantled, the source being left unprotected. Over the decades, since its first development, health considerations have made it increasingly difficult to use the neutron probe. Because of these limitations and also because of the difficulty of measuring the top 20 cm of soil with a neutron probe, alternatives were sought and developed. Nevertheless, the neutron probe continues to be used, where possible, because it is such an excellent instrument.

The capacitance probe

Techniques to determine the water content of soil by measuring its dielectric constant had been investigated before the neutron method was developed

(Debye 1929), and these became the basis for new instruments in several guises, one being the capacitance probe.

Principle   The *relative permittivity*, ε_r, of a material (also known as its *dielectric constant*) can be defined as the factor by which capacitance is increased over the free-space value (or free-air value, for all practical considerations). Since the value of ε_r for water is about 80 (it varies from 81 to 78 over the temperature range 15 °C to 25 °C), while for air it is 1 and for soil between 4 and 6, the capacitance value of a capacitor with soil as its dielectric is decided primarily by the amount of moisture it contains.

   However, care is needed with regard to how this capacitance is measured, because ε_r for many substances, including water, is not a constant but frequency dependent, having different values as the frequency rises from sub-audio up through the audio and radio bands to microwave. The reasons for this are complex and do not need to be fully explored here. Superficially, they involve the dipole nature of the water molecule and its tendency to rotate so as to line up with an electric field applied across it, such as that between the plates of a capacitor, and to reverse its orientation when the field reverses, for example when an AC signal is applied across the plates; the speed at which it does this (its *relaxation frequency*) is also a factor (Hasted 1973, Hoekstra & Delaney 1974, Smith-Rose 1933).

   As the amount of water in soil reduces, it concentrates largely where the particles touch, owing to capillarity, being perhaps only one molecule thick over surfaces not in contact. Under such conditions, bonding of the water molecules to the soil can be tight, restricting the movement of the water dipoles under the influence of an electric field. This affects the capacitance in an uncertain way, particularly when measured with a low frequency (tens of kHz). To minimise the problem, the frequency at which the capacitance of soil containing moisture is measured must be greater than about 30 MHz. Early attempts probably failed because audio frequencies were used (Anderson 1943, De Plater 1955). Later designs (McPhun 1979) used frequencies in excess of 30 MHz. By the 1980s, modern electronics and a better understanding of the physics involved had made the technique more practical, and portable capacitance probes became a reality.

Practice   Bell *et al.* (1987) and Dean *et al.* (1987) describe a probe based on a modified Clapp transistor oscillator running at about 150 MHz (in air), housed in a plastic tube of 4.4 cm diameter, by 29 cm long, with two concentric cylindrical capacitor plates (Fig. 9.5(*a*), (*b*)), the whole fitting into an access tube with an internal diameter of 5.0 cm. The access tube is similar to those

(*b*)

(*a*)

Figure 9.5. (*a*) A capacitance probe being placed in an access tube, with the hand-held frequency meter (*b*) connected by fibre optics to the electronics and the sensing electrodes in the lower (white) section of the probe. The grey upper tubing is used to locate the probe precisely at the required depth. (*b*) With the capacitance probe in place at the required depth, the frequency indicated on the meter is used to determine the moisture content of the soil immediately adjacent to the capacitor electrodes.

used for neutron probes, except that it is made of PVC to allow the passage of the electrical field into the soil. The probe is kept centred in the access tube by a soft nylon fabric ring at each end. The sensor tube can be moved up and down the access tube, in this case by attachment tubes rather than by cable. At each required depth, a reading of the frequency of the oscillator is taken by a hand-held frequency meter (Fig. 9.5(*b*)) at the surface; it is connected to the probe by a fibre-optic cable rather than by copper wire, so as to prevent stray capacitance from influencing the frequency of the oscillator.

More recently, a probe has been developed with two short rod electrodes that can be inserted directly into the soil for the measurement of the top 10 or so centimetres, without the need for an access tube (Robinson & Dean 1993; see Fig. 9.6). In this, it is similar to the *time domain reflectrometry* instruments described below.

Because of the dipolar nature of water molecules, it may be that the capacitance probe senses well the water that is free enough to respond easily to the reversals of the electric field, because it is not too tightly bonded, and less well to that more strongly bound by surface tension in the pore spaces and as thin films on particles, especially in the case of clay. Because of this uncertainty, there may be no simple relationship between the capacitance probe frequency and the MVF as measured by the gravimetric method or by the neutron probe method, each of which responds differently to the several types of water bond in the soil. There is probably no way of assessing with certainty which water molecules are sensed in the capacitance probe method. Nevertheless some reference is necessary for calibration, and the gravimetric method is the one used, because it is the most fundamental.

The bulk of the capacitance probe's response is derived from a volume of soil between 4 to 8 cm in thickness with a diameter of about 13 cm, the shape being approximately a vertical ellipsoid, which is much smaller than the sphere of influence of the neutron probe. While this may have the advantage of giving greater depth resolution, it also means that small anomalies in the soil can produce greater errors (Whalley *et al.* 1992).

Installation   Extreme care is necessary when auguring the holes and installing the plastic access tube so as not to allow an air gap to form between the tube and the soil, since this would dominate the frequency response. A perfectly installed access tube is essential and a gap as small as 0.5 mm affects the response; one or two millimetres is unacceptable. To achieve this standard of installation, special tools are needed and this puts up costs as well as demanding skilled and careful work. Air gaps can also be a problem when trying to work with shrinking clays, as the gap may form later, some time after a perfect installation. In the case of the

Figure 9.6. An important advantage of the capacitance probe is its ability to measure the upper few centimetres of the soil – for which the neutron probe is less well suited. To take advantage of this, a hand-held probe with short, external electrodes (Fig. 9.8) is an ideal tool for such measurements, being easily inserted and quickly read and removed.

hand-held probe, in which the electrodes are inserted directly into the ground, the problem of an air gap is less, but not absent.

Calibration　A capacitance probe typically operates in the frequency range of 80 MHz in water to 150 MHz in air. (In Fig. 9.5(*b*), the meter indicates 13437. This is the frequency in Hz divided by 8000, meaning that the probe output is actually 107 MHz.) This reading is converted into the relative permittivity, ε_r, using an empirically derived equation of the form

$$\varepsilon_{\mathrm{r}} = a/(R^2 - b) - c, \qquad\qquad (1)$$

where R is the reading and a, b and c are constants specific to each probe design and to each individual probe of any one design (be it the direct-insertion type or one that is lowered down an access tube).

Conversion from ε_{r} to MVF is, however, dependent on soil type and, for best results, the probe should be calibrated in the soil of interest. The much smaller volume of soil sampled by a capacitance probe means that the technique of calibrating a neutron probe, by taking soil samples from somewhere in the vicinity of a temporary access tube, is not applicable – since the scatter of points is too great. Instead, known-volume samples have to be collected as the actual access tube that is to be used is installed. This requires care so as to ensure that the samples, and the readings taken by the probe, refer to the same depth down the tube. Even this can give a spread of points because the probe is still not measuring the actual sample. The samples' water contents are then measured by the gravimetric method to determine the MVF and plotted against the probe's frequency readings; the MDWF can also be calculated from the soil samples.

The procedure gives only one point on the calibration curve – the MVF at the time of installation. To get around this, several access tubes, close to the original tube, have to be installed at different degrees of soil water content and the calibration process repeated each time; this method assumes lateral homogeneity of the soil, which may not be the case. Graphs produced in this way are approximately linear over the restricted range of soil moisture changes normally associated with each particular soil type, but graphs from different soils can have very different slopes and intercepts. If the various graphs from different soils are combined over the full range of water content, the overall relationship is non-linear, and some soils may not fit neatly onto this combination graph (see Fig. 9.7). In the case of a hand-held probe a similar procedure is necessary, the soil to be measured by the probe being removed for measurement by the gravimetric method. If this lengthy procedure is not possible, or pending its being done, a rough estimate of the MVF can be obtained by using another empirical equation; in the case of the Institute of Hydrology probe, and for typical UK loams and brown-earth types of soil, the equation takes the form

$$\mathrm{MVF} = 0.0977\,\sqrt{\varepsilon_{\mathrm{r}}} - 0.167 \qquad\qquad (2)$$

(Gardner *et al.* 1998, Robinson *et al.* 1998).

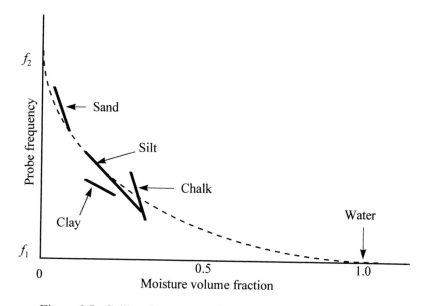

Figure 9.7. Calibration curves of a capacitance probe for different soils may be nearly linear, but each may have a very different slope and intercept, as this combination graph shows. The frequencies f_1 and f_2 will depend on the make of probe, falling roughly into the region 100 to 150 MHz. Thus, unlike the neutron probe with its linear calibration, the capacitance probe is far from linear. Further, as soils may often be layered, with one type overlying another, or mixed, the situation is further complicated.

The overall procedure, therefore, is to take a probe reading at the field site, convert it to ε_r by eq. (1) and then to convert this to the MVF by curves obtained over a period from field calibration or by using eq. (2).

Effect of soil conductivity An *equivalent circuit* of soil consists of a capacitor in parallel with a resistor, the resistor representing solutes in the moisture. It is necessary for the capacitance-measuring circuits to be able to cope with this shunt resistance, which can be quite low in some soils. For capacitance methods to be successful, the effect of the conductivity of the soil must be understood. When the shunt resistance is high, the capacitance measured by the electrodes is correct, but as the resistance falls the oscillator response is damped, the oscillator frequency reduced, increasing the apparent ε_r and so giving an overestimate of the water content. If the conductivity is high (2000 $\mu S\,cm^{-1}$, equivalent to a salt solution of 1000 ppm – see Chapter 10), ε_r can be overestimated by as much as 30 (Robinson *et al.* 1994). For such soils, a relationship existing between the true capacitance, the apparent capacitance, the resistance and the angular frequency (the frequency in hertz multiplied by 2π) can be used to derive the correct capacitance (Dean 1994), although work remains to be done on the effects of temperature on the conductivity. However, with a conductivity below about 500 μS cm^{-1} there is little effect, and most soils are not conducting enough to cause problems, although some clays and saline or fertilised soils can put the measurement badly in error.

Comparison of the capacitance probe with the neutron probe   The care required in the installation of access tubes in order to avoid air gaps, compared with the relative ease of neutron-probe access-tube installation, is a considerable disadvantage. This, taken along with the need to calibrate the capacitance probe's response in different soils for precise work, the uncertainty over exactly which water molecules the instrument responds to and the much smaller volume of soil measured, considerably offsets the method's originally perceived advantages over the neutron probe, of low cost and simplicity. Neither advantage has yet materialised, the cost of a capacitance probe being not very different from that of a neutron probe; both are priced at between £5000 and £7000, in part owing to the equipment required to instal the access tubes accurately. Indeed the neutron probe, despite its main disadvantage of using a radioactive source, must still be seen as the reference, the more accurate and preferred method (Evett & Steiner 1995). But with research and development continuing, the capacitance method is likely to become more sophisticated, or perhaps it will be superseded by the time domain reflectrometry method of measuring a soil's dielectric constant, which we now discuss.

Time domain reflectrometry

Time domain reflectrometry (TDR) is another way of measuring the dielectric constant of soil – and thereby its moisture content. It is not a new concept; references to its application to soil moisture measurement go back over the last three decades (Hoekstra & Delaney 1974, Davis & Chudobiac 1975, Topp *et al.* 1980, Dasberg & Dalton 1985, Nadler *et al.* 1991, Whalley 1993).

Principle   A fast rise-time (200 picosecond) electromagnetic step pulse, launched down a transmission line (such as a coaxial cable or twin feeder), travels not at the speed of light but at something less, the actual speed depending on the value of the relative permittivity ε_r of the material filling the space between the conductors. The larger the value of ε_r, the slower the speed of travel. Because the value of ε_r for water is about 80 while for soil it is less than 6, the speed of travel of a pulse down a transmission line with soil as its dielectric is largely dictated by the soil moisture content (the MVF). To be precise, the combination of soil, air, water and roots has what is known as an *apparent dielectric constant* because it is a mixture of dielectric materials. The TDR method operates by measuring the time of propagation of the pulse along the line to the end, where it meets an open circuit; this causes the energy to be reflected back up the line, returning to its origin. (If the line were terminated in a load, such as a matching impedance of the same value as the source and the transmission line, the energy would be absorbed and there would be no reflection.) It is the round-trip time of the pulse that is measured, an average of perhaps 1000 readings being taken by the instrument. Times as short as 2 to 3 ps (10^{-12} s) have to be measured with short transmission lines in dry soil, but typically the round-trip time is several nanoseconds. In addition to a reflection from the end of the line, if there are changes in ε_r along the length of the transmission line then the consequent change in impedance causes a fraction of the signal energy to be reflected at each discontinuity back up the line. By an analysis of the resultant waveform, further information can be obtained about the changes in the moisture content profile, in addition to that obtained by simply measuring the round-trip time for the pulse.

For laboratory experiments investigating the TDR method, it has been more convenient to fill a coaxial transmission line with soil, but for field use a parallel transmission line, consisting of two or more parallel metal rods similar to the capacitance probe (Fig. 9.8), is more practical and the most widely used. A coaxial cable with a characteristic impedance, Z_0, of 50 Ω is the most convenient type of transmission line to interconnect between the pulse generator and the parallel transmission line in the ground. But the two cannot be connected together directly, since a parallel transmission line has a value of impedance different from that of a coaxial feeder. A coaxial cable is unbalanced; its inner cable carries the signal and its outer screen is 'earthed', the field being contained within the screen. A parallel transmission line is balanced, both lines carrying the signal, the field being between and around the conductors. At the point where the balanced meets the unbalanced line, they must be interconnected through an impedance-matching transformer (known as a *balun* – an abbreviation for balanced–unbalanced). This applies to those sensors that are made up of two parallel conductors (Fig. 9.8).

There is an alternative design of sensor based on three (or more) conductors and, as this acts as an unbalanced transmission line like the coaxial cable, no balun is then needed; the inner conductor is connected to the inner coaxial conductor and the outer conductors to the coaxial screen (Fig. 9.9). The sphere of influence is roughly the volume of an ellipsoidal cylinder with a greater diameter of about 8 cm and a length of 5 to 15 cm, which is akin to the capacitance probe's (and considerably less than that of the neutron probe), with the same problems that this causes for the capacitance probe.

Figure 9.8. In common with both the capacitance probe and TDR for near-surface measurements, the sensing electrodes take the form of two parallel rod electrodes from 5 to 15 centimetres in length. The ones illustrated are those of the hand-held capacitance probe of Fig. 9.6.

If the soil moisture contains dissolved salts then although the amplitude of the signal is reduced the travel time of the pulse is not affected, but there is a limit when the conductivity rises towards 2000 μS cm^{-1}, because the waveform becomes so attenuated and smoothed that it is difficult to detect the pulses reliably. But, if not too large, the attenuation is not entirely a disadvantage since it is possible to use it to measure the conductivity of the sample.

Practice TDR is not the exclusive preserve of soil moisture measurement; indeed, it is a specialised application of it, a more general use being the detection of faults in long cables. For convenience, some workers have simply used cable testers for the measurement of soil moisture and these have built-in oscilloscopes which aid interpretation of the trace, but today most TDR systems, with and without oscilloscope displays, are custom built for measuring soil moisture and there are now several makes of instrument available. Some simply measure the transit time of the pulse and convert this into ε_r and thus into MVF; others also include an oscilloscope that allows the reflected waveform to be displayed and thus more information to be gleaned from the reflections due to changes in ε_r at different depths, as mentioned earlier.

At its simplest, a measurement is made manually by pushing the sensor into the ground (as shown in Fig. 9.6 for the capacitance probe), taking a reading on a built-in meter, withdrawing the sensor and moving on to the next site. If measurements were taken of the same volume of soil each time, then reinsertion of the probes into the same holes could give rise to air gaps around the probe rods. This procedure might also allow more rainwater to enter that area (via the holes left after taking a reading) than the untouched ground nearby. To avoid these problems, and yet also guard against spatial variation, several readings are usually taken at different nearby points and an average produced.

Manual insertion of a probe into the ground only provides a reading for the top 5 to 15 centimetres of the soil. If deeper measurements are required, one option is to bury a sensor permanently, a coaxial cable connecting it to the surface where either a manual reading can be taken with a hand-held meter or the measurement can be logged automatically. If several probes are installed permanently, then they are connected by cables to a multiplexing switch that allows them to be scanned and their readings logged.

If a profile of soil moisture is required, one option is to bury a sensor at each required depth but, at around £300 a probe, this can be expensive and, where probes contain the electronics built-in, the price can be four times as much as this. (Where the electronics is separately contained in a hand-held portable unit with a display, its cost is in the region of £4000; with an oscilloscope it is nearer £8000.) The presence of a buried probe can also affect the behaviour of the soil being measured. There may also be a period of stabilisation after installation. Nor is it possible to obtain a profile at precisely one point, since if several probes are to be installed at different depths, they must be displaced laterally from each other. This can introduce uncertainty unless the soil is homogeneous. If a pit is dug adjacent to the point where the profile is required, probes can be inserted sideways into the ground from the pit, immediately over each other. But this has its disadvantages, since it disturbs the ground even more than augering access holes from the surface.

To avoid these problems, an alternative to burying several probes at one site is available in the form of a probe that can be moved up and down an access tube, as with the capacitance probe (Fig. 9.5(*a*), (*b*)). This probe is constructed with its transmission line in the form of two vertical metal strips on opposite sides of the probe, much as the capacitance probe has two horizontal rings as capacitor plates. However, because the plan view of the sphere of influence of the vertical plates is elliptical (Fig. 9.9), it is necessary to take several readings at each depth, rotating the sensor so as to give all-round coverage. This is also true for the portable probes with external rods that are simply pushed into the soil by hand, a plan view of their sphere of influence also being an ellipse.

Furthermore, just as it is essential to ensure that there are no air gaps between the access tubes and the soil when installing a capacitance probe, it is equally important with TDR sensors; the sensitivity decreases exponentially into the medium so that the volume nearest to the tube has greatest influence. Similar special equipment is, therefore, advisable to enable the (fibreglass) access tubes to be installed with minimal air gaps.

Calibration   Whereas each capacitance probe requires a calibration of its frequency against ε_r to be obtained, with TDR this is not necessary since the travel time of the pulse is related directly to ε_r. However, the conversion from ε_r to MVF requires exactly the same procedure as that described for the capacitance probe (Robinson *et al.* 1998).

Modified TDR   A modification to the TDR method (which we might dub CW TDR) has been developed in which a continuous wave of about 100 MHz is emitted, instead of a single fast-rise pulse (Gaskin & Miller 1996). The continuous wave is reflected from the end of the line just as is the pulse. The outgoing and returning signals interfere, producing a composite standing wave, the amplitude of which depends on the phase and amplitude of the reflected wave (which either adds to or subtracts from the source wave). This in turn depends on the length of the probe and the soil's relative permittivity, and thus its moisture content. The ratio of the outgoing wave to the composite standing wave is calibrated against MVF, using gravimetric techniques similar to those developed for the capacitance and TDR instruments. The advantage of the method is that it does not rely on the detection of pulse thresholds, which can be somewhat uncertain, nor on measuring the very short travel times of pulses. In consequence it is cheaper. Otherwise it is similar to TDR, with the same advantages and disadvantages.

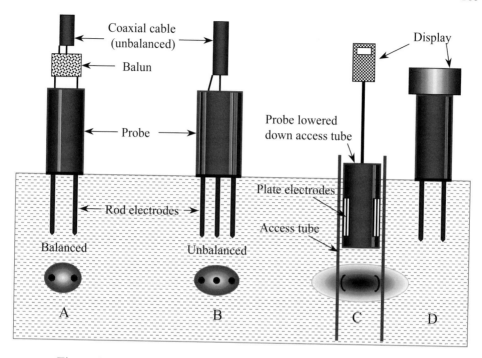

Figure 9.9. TDR probes can take several forms. In A, two short rods are inserted into the soil (or the whole can be buried in the ground at any required depth). Because the probe's two electrodes form an electrically 'balanced' system, a balun must be interposed between the probe and the 'unbalanced' coaxial cable transmission line to match their impedances. The coaxial cable can run some distance to a data logger. B is similar except that the probe's three electrodes form an unbalanced system and so can be connected directly to a coaxial cable. In C, a supporting cable allows movement of the probe up and down an access tube, as with the capacitance probe (Fig. 9.5). However, in this case, because its field of influence is an ellipsoid, the probe must be turned several times at each depth to get a fully representative reading (whereas the capacitance probe uses two ring electrodes one above the other, resulting in an equal response all round). D is the same as A, except that it contains the display built into the top, making it a portable system similar to the capacitance probe of Fig. 9.6. In cases A, B and C the measured volume, an ellipsoid, is shown below the electrodes.

There seems to be some uncertainty whether to call this method TDR or a capacitance probe. Indeed workers in the field are hesitant to pigeonhole it, saying that in reality both capacitance probes and TDR overlap in principle. This is certainly true in that both measure the dielectric constant of the soil, albeit in different ways.

TDR compared with the capacitance and neutron probes   Because the TDR method measures the soil's ε_r, it is subject to many of the same limitations as the capacitance probe. Air gaps around the access tubes introduce errors. There is the same uncertainty as to which water molecules are measured (the question of their levels of binding and water of crystallisation), although this is less of a problem if it is changes in water content that matter. The same very careful and lengthy methods are required in calibrating a TDR probe against readings obtained using the gravimetric method (for precise scientific work, that is, although perhaps not for less demanding applications). The relatively small volume of soil sampled by the probes (their sphere of influence) makes the method more vulnerable than the neutron probe to local anomalies within the soil such as stones and other (unknowable) spatial variations.

In consequence, for applications such as estimating irrigation needs, which do not require precise quantitative measurements, for measurements near to the surface and when automatic logging is necessary, dielectric methods have much in their favour. But for accurate, precisely calibrated data, collected manually, the neutron probe is still difficult to better.

By radio propagation

Yet another way of measuring the dielectric constant of soil, and so its moisture content, is to measure the effect of soil moisture on the passage of a radio wave (Hasted 1973).

When a transmitter launches a radio wave towards a receiver some distance away, the received signal strength is dependent on the relative permittivity of the ground between the two antennae; the presence of soil moisture increases the field strength in direct proportion to the MVF, linearity being maintained over a wide range of moisture content. The effect can be visualised as being like that of a large capacitor, the antennae being the plates, the ground the dielectric. Vertical polarisation is best, since a horizontally polarised wave is very dependent on antenna elevation. The main problem with vertical polarisation is that the signal is attenuated by thick green vegetation, although low-growing plants such as grass, or sparse vegetation, have little effect. Because radio waves are attenuated exponentially with depth into the ground, it is the top metre or so of ground that has most effect on signal strength.

The method has been tested successfully at frequencies ranging from 30 to 150 MHz. There must be a minimum of eight wavelengths between the transmitter and receiver antennae, allowing distances of from 15 to 300 m or more to be measured (Chadwick 1973); this gives an integrated value of soil moisture over a large area, unlike the other methods, which give localised, spot measurements on small volumes of soil at specific depths. Perhaps because of its relatively broad-brush picture, the method has not been exploited in the way that the methods that produce spot readings have, although its large areal coverage could be seen as an advantage.

By gamma ray attenuation

Just as snow can be measured by its attenuating affect on gamma rays, so too can soil moisture. The principle is identical: a source of gamma rays is placed in a tube below the surface, a Geiger–Müller detector being sited some distance away to measure the level of gamma rays transmitted through the soil. Scattering and absorption of the rays occur in their journey from source to receiver, and although the soil is the principle determinant of these processes, variations in gamma ray intensity over time are due largely to changes in water content, provided that the soil is not subject to shrink–swell behaviour.

The method is little used for soil moisture measurement, the other options generally being preferred; however, it is more widely used for measuring soil bulk density (see the gravimetric method above) in civil engineering applications. For this type of measurement, the source (typically caesium 137 with a strength of 10 mCi) is housed in a portable unit from which it can be lowered incrementally to a depth of about 30 cm, the detector being housed in the base of the unit. An alternative operation is to sense the back-scatter of the gamma rays, as Fig. 9.10 shows, removing the need to prepare an access hole.

By heat conduction

The rate of heat dissipation in soil is a function of the soil's thermal conductivity, which increases with water content. It is thus possible to measure moisture content by measuring the rate at which an electrical heat source, buried in the soil, warms up – the greater the water content the slower the rate. The idea was first proposed in the 1930s (Shaw & Baver 1939), but because of the problem of maintaining good thermal contact between the sensor and the soil, it has not found the wide use that the other methods have. If it could be made reliable it would be a low-cost and simple method, requiring only a heat source and a means of measuring temperature (Bloodworth & Page 1957).

Measuring soil tension

As the level of soil moisture changes, so does its tension – the dryer the soil the higher the tension. The significance of a combined measurement of soil moisture and soil tension is that not only is the amount of water known but its direction of flow can also be inferred – up, in the case of evapotranspiration, and down in the case of groundwater recharge. Soil tension can be measured in two main ways, by tensiometer or by gypsum resistance block, as follows.

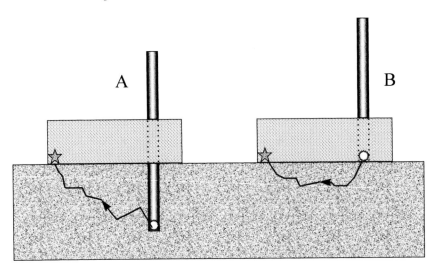

Figure 9.10. In A, the gamma-ray path is directly from source (white sun-burst) to detector (shaded star); in B the path is by back-scatter. The former requires an access hole to be made, although this gives control over the depth to which the soil moisture is measured. In the latter case, the instrument can be stood simply on the ground, although only the top surface is then sensed.

Porous-pot tensiometers

Manually read tensiometers   A porous pot, filled with water, is connected by a plastic tube, also filled with water, to a mercury manometer (Fig. 9.11(*a*), (*b*)). A stopper at the top of the tube to which the porous pot is fixed has a smaller bung, which is used to allow the system to be filled with water and purged of air. If the pot is held in air and at the same level as the mercury, the mercury level in the tube remains at reservoir level. If the pot is lowered, mercury is drawn up the manometer tube by an amount equal to the pressure-head (suction) caused by gravity. This suction due to gravity increases the further the pot is lowered. This also applies when the pot is in the soil, and so needs to be subtracted from any suction caused by the soil. Since it is entirely a matter of height difference, it can be calculated.

If the pot is now installed in the soil, soil tension adds its negative pressure to that of gravity and so draws the mercury further up the manometer tube, by an amount additional to that due to gravity – unless the soil is saturated, when there is no soil tension. The installation procedure, to ensure good hydraulic contact with the soil, is the same as the procedure, explained later, for gypsum blocks. Several pots can be installed at one site at different depths, usually corresponding with those at which the MVP is measured, each with its individual manometer tube (Fig. 9.3). Manometer readings are taken at the

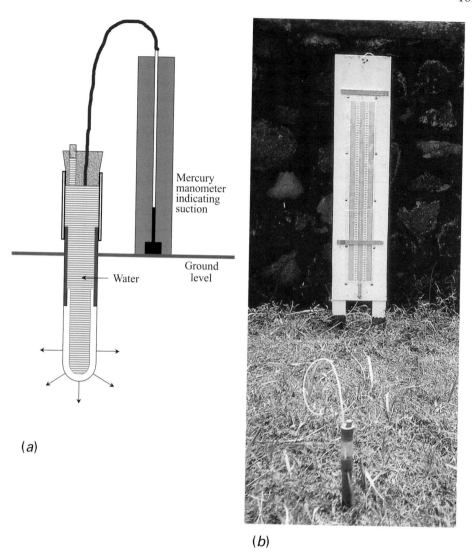

(a)

(b)

Figure 9.11. (*a*) The unglazed ceramic U-shaped porous pot of a tensiometer
is fixed to a plastic tube (dark shading), the top few centimetres being made of
transparent material to allow the water inside it to be seen. The top of the tube
is sealed with a rubber bung, through which passes a flexible small-bore
plastic tube, connected at its far end to a mercury manometer. The small bung
is removable to allow the system to be purged of air, using a syringe. The
arrows below the porous pot show the effect of tension plus gravity on the
water. (*b*) With the porous pot installed in the ground, the negative pressure
produced by soil tension and gravity draws mercury up the manometer tube,
which is read manually at the required interval, often at the same time as the
moisture content is read with, for example, a neutron probe (Fig. 9.3). The top
of the tensiometer is seen in the lower foreground and the flexible tube
connecting it to the manometer scale board in the background.

same time as the MVP is measured. This might be daily, weekly or monthly, depending on the conditions and requirements.

Manual tensiometers are also manufactured in which the mercury manometer is replaced by a mechanical vacuum dial gauge, with the means to fill the system with water and to ensure that no air is present in it. These are useful for obtaining rapid readings in the top soil layers during a field visit; they are not left *in situ*.

Unfortunately, this type of tensiometer can only function over a small range of tensions, from 0 to about −0.8 bar, because, as the vacuum pressure increases, air bubbles start to form in the water, drawn out from crevices or from the pot itself. (The water used should be boiled to remove the air dissolved in it in order to lessen the effect, but this does not fully prevent it.)

Electronically measured tensiometers   It is possible to connect an electronic pressure sensor (see also Chapter 7, the section on electronic barometers, and Chapter 10, the subsection on sensing water level by pressure) to the same type of porous pot as is used in the manual tensiometer, and so to measure the pressure automatically (Fig. 9.12(*a*), (*b*)). The engineering of this type of instrument makes it much more complex than a manual tensiometer and therefore much more expensive. And again these tensiometers can only operate up to about 1 bar suction.

Gypsum blocks

While most of the water useful to plants is in the region covered by porous-pot tensiometers (up to −0.8 bar), it is important to be able to measure well beyond this range because the processes of evaporation, transpiration and the downward percolation of water continue at higher tensions.

Principle   The electrical conductivity of soil varies with its water content because the water contains dissolved salts, making it much more conductive than the soil particles. But because of the highly variable nature of soil, it is better to measure the conductivity not of the soil itself but of a porous block kept in close contact with the soil. The water content of such a block tends to be in equilibrium with the surrounding soil and so measurement of its resistance provides a measure of soil tension. Such sensors are known as *resistance block tensiometers*. Variations in water content of the block result also in a capacitance change but it is usually the resistance that is measured because it changes more than the capacitance in response to soil water potential, from around 500 Ω at −0.4 bar to 10 000 Ω at −10 bar.

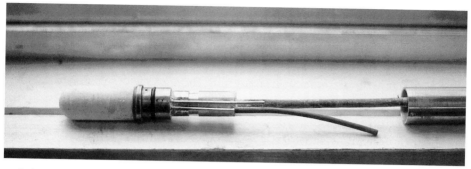

(*a*)

(*b*)

Figure 9.12 (*a*) The negative pressure sensed by porous pots can also be measured electronically using a similar type of pressure sensor to that used to measure water level (Figs. 10.10, 10.11); it is seen in the figure affixed immediately to the right of the porous pot. (*b*) At the surface, automatic tensiometers are filled with water and purged of air in a similar way to the manual type, but with valves rather than a simple bung. The cable from the sensor carries the signal to a data logger, and usually several such tensiometers are operated together to give a depth profile. Within the cluster of tensiometers shown (on the right) is a neutron probe access tube, allowing the amount of moisture to be apportioned, upward to evaporation and downward to percolation.

Because the conductivity of soil also depends on the quantity of dissolved material it contains (as well as on how much water it contains), it is desirable to minimise the effect of the dissolved material so that just the quantity of water is sensed when the resistance is measured. While soil tension resistance blocks have been made of various porous materials, those made of gypsum (or plaster of paris – the commonly occurring mineral hydrated calcium sulphate) are preferred because the water they contain is always saturated with calcium sulphate. In non-saline soils, the concentration of natural salts is likely to be much less than this and so using gypsum buffers the conductivity of the water they contain against variations in the soil's conductivity. (In acidic soils the blocks quickly dissolve.)

Since the 1940s, gypsum blocks have been used to measure both soil moisture and soil tension but because of the confusion that arose between the two uses, they fell into disfavour in the 1960s and 70s and were looked at with suspicion until the situation was clarified (Wellings *et al.* 1985). Resistance blocks are unreliable as soil moisture content sensors, owing to the hysteresis of the relationship between soil moisture content and the potential of the soil and that of the block itself. But if blocks are individually calibrated on a drying cycle and are used for tension measurements only on a drying cycle, good results can be obtained. On a wetting cycle results are only approximate. Calibration is done in a ceramic pressure-plate apparatus, the blocks being covered in a thick slurry of, for example, chalk. Instead of applying suction to the outside of the ceramic plate, pressure is applied to the inside, this having the same effect (Wellings *et al.* 1985). It is necessary to put blocks through several wetting and drying cycles before calibration, however, as their resistance characteristics change considerably at first. Figure 9.13 illustrates a typical resistance versus pressure curve of a gypsum block calibrated in this way. Because calibration by this method takes several days, just three points are measured as an adequate compromise.

Construction and installation   Gypsum blocks are probably the simplest of all of the sensors described in this book, although manufacturers may challenge this and point to the many pitfalls of do-it-yourself construction. They are quite right, since they are cheap to buy and consistency from one to another is much more certain in the bought-in item. All designs consist of plaster of paris moulded around two electrodes. While some are flat, a common design is cylindrical, with concentric stainless steel mesh electrodes (Fig. 9.14).

Installation is similar to the installation of a neutron probe access tube, by auger. When the required depth is reached the block (previously well soaked in water) is lowered to the bottom of the tube and pressed into good contact with

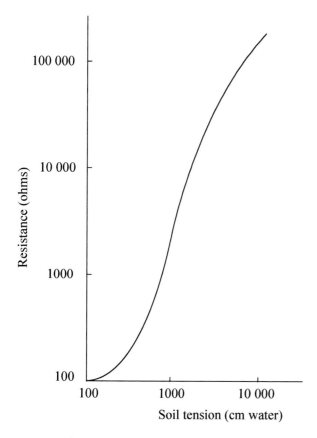

Figure 9.13. A gypsum block of the cylindrical type shown in Fig. 9.14 will have a response roughly of the shape shown here. Other designs, such as those with parallel plates, have slightly different responses.

the ground. Thereafter the hole is back-filled with soil slurry for about 50 cm, which is consolidated by gentle pressure with a rod or tube. The remainder of the hole, if it was deeper than 50 cm, is filled with dry soil sieved to pass a 2 mm mesh. It is difficult to retrieve blocks at a later date if they are much below 50 cm, the only way being to dig them out, but as they are low cost this is not essential.

Accuracy and corrections   These are not precision sensors. If semiquan-titative data are all that is required then resistance blocks can be used without calibration, but if more is needed the calibration briefly described above must be used (Wellings *et al.* 1985). In addition to calibration, temperature correc-tion is needed and for this thermistors can be installed along with the blocks to monitor the temperature at the time the blocks are read; errors are in the order

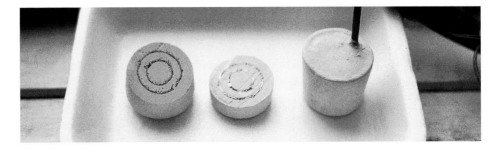

Figure 9.14. Much simpler and cheaper than porous pots, and with certain technical advantages (a much greater tension range), gypsum blocks are constructed of (typically, but not exclusively) two concentric stainless steel mesh cylinders embedded in gypsum, here sectioned to show the interior construction.

of 1.5% K^{-1}, although temperature changes are slow and small below ground. Because the sensors can change through chemical reaction with the soil, their calibration will drift with time. This can be detected by comparing data a year apart, at similar soil water contents, and this allows some correction to be applied. Depending on the chemistry of the soil, blocks will last for three to five years. While the absolute values of matrix potential may be in error by $\pm 20\%$ after a year or more, if only the depth of the zero-flux plane (ZFP) is required this may be acceptable. Response times vary from minutes at low tensions to hours at 5 bar and up; this is adequate to follow the changes that occur in natural soil conditions.

Measuring block resistance   A DC bridge or digital multimeter is not suitable for measuring the resistance of gypsum blocks because electrolysis occurs and the blocks polarise, causing a rapid rise in resistance due to gas bubbles forming on the electrodes. Bridges for manual use operate at about 1.5 kHz, the operator balancing the bridge to a null point on a meter by hand. For automatic logging, various methods can be used to prevent polarisation. In one design (Strangeways 1983) a single, short (120 ms) DC pulse is applied to the block via a resistor, the voltage across the resistor being logged. The block is then short circuited until the next reading to discharge any slight polarisation that has occurred. In all methods, the block's resistance must be measured in isolation from the other blocks of a profile since there is a resistance path through the soil between blocks that will otherwise affect the readings.

Accessing groundwater

Since instruments for ground- and surface-water measurement, with respect to both depth and quality, are similar in principle, they are most logically dealt with together in the next chapter, on surface water measurement. Here we look at how access is gained by drilling to the groundwater, in order to measure it.

Drilling methods can be divided into rotary and percussive, either of which may or may not use circulating fluids (Nielsen 1991).

Rotary drilling

Hand augering   Hand augers offer a simple method of drilling holes in the top few metres of ground provided it is of unconsolidated material. Several types are available, from the familiar helical auger to those for soft fine soils and for gravel; the cutting tool is mounted on a rod, the operator pressing down while turning a handle until the auger is full, when it is withdrawn. Holes of 5 to 10 metres depth or more can be augered in this way, with diameters up to 20 centimetres.

Continuous-flight and hollow-stem augers   A central shaft about 1.5 metres long, known as a *flight*, has a steel helix wound along its full length. Flights can be joined together, forming a continuous helix that acts as a conveyor belt to the surface (Fig. 9.15), torque being transmitted from the surface through the flights. The auger's cutting head makes a hole 10% wider than the diameter of the helix, to ease withdrawal and to reduce wear and friction. Such augers are generally used to drill unconsolidated formations, although rock can some-times be penetrated if the appropriate bit is used. Diameters range from 10 to 60 cm with depths in excess of 25 metres.

In *hollow-stem* augers, the rod up the centre of the helix is replaced by a tube, the drill bit being in two parts, one on the helix and one on a rod that passes down the centre of the tube, both rotating together. (Alternatively the central bit can be run down the stem on a wire cable and locked onto the auger.) The advantage of hollow-stem augers is that the central rod, or wire cable, can be withdrawn, leaving the tube and helix in place; this allows a water sampler, measuring instrument or casing (see later) to be inserted. It also allows a change in drilling technique to be made at an interface between uncon-solidated formations and rock. Cutting speeds vary from 30 rpm for heavy clay to 75 rpm for sand (Clayton *et al.* 1982).

Figure 9.15. One of two principal ways of making a borehole for monitoring groundwater is to drill the hole using a *continuous-flight* auger, which acts as a conveyor belt to the surface. In this case the auger is of the *hollow-stem* type, allowing access to the cutting surface down its centre, perhaps to take water samples or so that casing can be inserted. Torque is generated by the hydraulic equipment at the top. Several flights can be joined to reach greater depths.

Percussive drilling

Here the method of advance is by hammering, striking or beating the formation. There may also be some rotation, but this will be mostly to maintain the roundness and straightness of the borehole.

Driven wells   The simplest way to instal a shallow, small-diameter borehole in unconsolidated formations free of large stones is to drive a *well point* (metal cone) into the ground using a pneumatic or hydraulic jackhammer or a sledge hammer, a tube behind the point containing holes (a *screen* – see later) to allow water to pass into it from the surrounding formation. To prevent damage to the screen during installation, a rugged outer tube (or casing – see later) can be used to carry the driving force to the tip, the casing being partly or completely withdrawn on completion of installation, in order to expose the screen to the water. Alternatively, the driving force can be applied directly to the point using a down-hole weight, raised and lowered on a rope.

Cable tool drilling   The earliest drilling rigs were probably made about 3000 years ago in China using bamboo. In the present-day equivalent, the rig is of steel and the *tool string* is raised and lowered at a short distance from the bottom of the borehole, breaking off fragments of rock or loosening unconsolidated material. The string comprises a drill bit, a heavy length of steel (the *drill stem*) to give the necessary mass to break the material (and also to keep the hole vertical), a *jar* to help if the tool string becomes jammed (by back-hammering) and a *swivel socket* that allows the string and thus the bit to rotate at random by virtue of the lie of the *steel wire cable* (or *rope*) on which the tool string is suspended. The *cable* is moved up and down by a *spudding arm* fixed to the rig.

Various bits are used depending on the formations being drilled, although they are mostly chisels of various types. To remove the cuttings, water is added (if the hole is above the water table) to form a slurry which is then periodically bailed out using a *shell* or *bailer* (see next section). For gravel, a bit with 'orange peel' jaws may be dropped through the gravel, the jaws closing as the bit is withdrawn. Temporary casing is inserted behind the bit to prevent collapse.

Although this is one of the oldest drilling methods, it is still widely used because it is simple and cheap and does not need large volumes of water. It is, however, slow (perhaps 5 to 10 metres a day) and holes smaller than about 15 cm diameter are not practicable because of the need to use relatively large and heavy bits and drill stem to make adequate progress. The method is, therefore, less used for drilling monitoring-boreholes than for production-boreholes, but there is some overlap between drilling techniques and borehole use (Nielson 1991).

Light percussive drilling   This type of drilling is also known as *shell and auger* (originally non-cohesive formations were shelled and cohesive formations were augered using this method). It is a scaled-down form of the cable tool method and is used mainly for unconsolidated formations (Fig. 9.16). In place of the spudding arm, the reciprocating motion is here imparted manually by using a clutch on a simple diesel engine, which raises the string and then lets it drop. The same cutting tools are used (with a *sinker bar* to give the necessary weight – compare with the *drill stem* of the cable tool method). When a clay cutter is used, the clay or soil usually sticks inside the cutter and can be drawn up and removed manually. With gravel and sand, a *shell* (or *bailer* or *sand pump*) is used. This is a heavy tube that is surged up and down about 30 cm each second, the slurry and suspended matter being forced by this action past a non-return Clack valve in the base of the tube, the tube then being drawn to the surface and emptied. The casing must be advanced while drilling to prevent collapse (Dixon 1998).

Drilling with circulated fluids

The drilling techniques described so far, both rotary and percussive, have not involved the use of fluids pumped down the boreholes. The following methods use circulated fluids – water, air or 'mud'.

The fluid is usually pumped down a central drill pipe, emerging out through the cutting bit and returning back up the annular space between the borehole wall and drill pipe. The purpose of the fluid is to cool and lubricate the bit and to carry cuttings back to the surface. To achieve the latter action, a certain minimum velocity is needed, depending on the fluid and on the particle size. For water this is $1 \, \mathrm{m \, s^{-1}}$, and for air $15 \, \mathrm{m \, s^{-1}}$. Because air is everywhere, and water may have to be transported, air is often the more convenient, but air compressors are expensive both to buy and to operate. However, air has the advantage that it does not contaminate the borehole to the same extent that water can, although even the compressed air will contain hydrocarbon lubricants.

A thin, mud-like, mixture of water and bentonite (a naturally occurring clay of hydrous aluminium silicate mineral) is sometimes used in the place of plain water, helping to raise particles more easily. A mixture of air and water, as foam, is another commonly used fluid. Alternatively polymers may be used as the 'mud', although these can affect the chemical make-up of the water in the

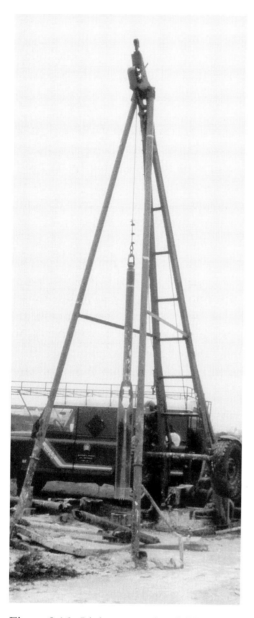

Figure 9.16. Light percussive drilling is a simple, if slow, way of producing a well. In the figure, the *tool string* is made up of, from top to bottom, a *swivel*, *sinker bar* and *clay cutter*.

borehole even more than bentonite can. Indeed the use of fluids to aid drilling is always liable to contaminate the borehole and its contents, whatever the fluid used, and this is not ideal when the measurement of the quality of the groundwater is the purpose of drilling the borehole. But it is often unavoidable, particularly for the deeper boreholes.

Direct-fluid circulation with rotary drilling   A drill pipe with an attached bit is rotated continuously against the face of the borehole while fluid is pumped down the pipe. The down-force required on the bit is typically between 200 and 400 kilograms force for each centimetre of the bit diameter, with rotary speeds of 30 to 250 rpm depending on the formation being drilled. The cutting tool may be a three-toothed cutter of hardened steel or tungsten carbide, which rotates on its own bearings (a *rock roller*), or for soft formations a three- or four-blade *drag bit*. To keep the bit straight, a heavy *drill collar* can be fitted above it. The drilling fluid is pumped into the top of the drilling tube and emerges through the drill bit, returning to the surface up the annular space between the hole wall and the central tube. Problems can occur, however, if the borehole diameter is large, since the pump may not be capable of producing enough updraft in the larger annular space between borehole and drilling tube, and so may fail to raise the cuttings. To overcome this problem, reverse circulation is used.

Reverse-fluid circulation with rotary drilling   In this method, the drilling fluid is simply allowed to fall by gravity down the annular space until the hole is full of water up to the surface. It is then pumped back up the drilling tube (in the reverse direction to the previous method). Because the diameter of the drilling tube, and thus the pump power required to achieve the minimum velocity and so lift the cuttings, is a constant, larger-diameter boreholes can be drilled. Although this method is not much used for drilling boreholes for monitoring applications, since large-diameter holes are not required, a modified form of it, known as *centre-stem recovery* or *dual-pipe drilling*, is useful for making monitoring boreholes. Here two concentric drill pipes are used, the fluid being pumped down the annular space between the two tubes to the bit, which cuts the hole in such a way as to deflect the fluid back up the central tube. The spoil is either collected as a sample or allowed to pile up on the ground.

Top hammer drilling   Also known as *drifter drilling*, this method combines percussion, rotation and circulating fluids and although it has been used to instal access tubes for neutron probes (see Chapter 9 on soil moisture) it is used mostly for drilling in mines and quarries for blasting. The hammer is a scaled-up pneumatic or hydraulic road-breaker, with the addition of rotation, all mounted on a rig. Air is the usual drilling fluid but water can be used. Penetration rates are high but diameters are limited.

Down-the-hole hammers   Having a pneumatic hammer immediately behind the bit avoids the necessity of a top hammer, when the impact energy has to be passed via drill rods or tubes, in which it will be partly dissipated. In a down-the-hole (DTH) hammer, blows are delivered to a tungsten carbide *button bit* at 500 to 1000 blows a minute, while the whole drill is rotated at 10 to 30 rpm. The exhausted air from the hammer cools the bit, and also brings the cuttings to the surface up the annulus between the drill and the borehole wall, the updraft being about 15 metres a second. The method works well in hard formations but is unsuitable for unconsolidated material.

Casing and screening

Having drilled the borehole, it must next be *developed*; this comprises a cleaning-up process to rid the borehole of sediments or contaminants introduced during drilling. It may be done in several ways but a typical procedure is to stir up the water at the base of the hole by raising and lowering a *surge block* on the end of the drill rod and then pumping the water out, the process being repeated until clear water is pumped up. Alternatively compressed air can be used to stir the water.

Unless the borehole is in rock, its walls will have to be supported by *casing-tube* inserted either during or after drilling the borehole, depending on the formation and the drilling method. In order to allow water to enter the well, part of the casing has to be in the form of *screening*: this is a tube containing fine slits (Fig. 9.17) or in the form of a metal mesh. The materials used for both casing and screening are either plastics such as PVC or ABS, or some form of steel or fibreglass-reinforced epoxy or plastic. It is not necessary here to go into the intricacies of how the casing and screening tubes are placed in the borehole, but four typical end-products are illustrated in Fig. 9.18.

To ensure a tight seal at the required depth, the casing and screen may be installed centrally in the borehole, leaving an annular space around them (Fig. 9.19). Below the level of the screen the annular space is filled with a sealing compound, usually the clay bentonite (see the discussion of drilling fluids), inserted as pellets down the annular space to the point required, where, in contact with water, it swells and forms a watertight seal. (However, chemists may object to the use of bentonite as a seal on the grounds that it can contaminate water samples.)

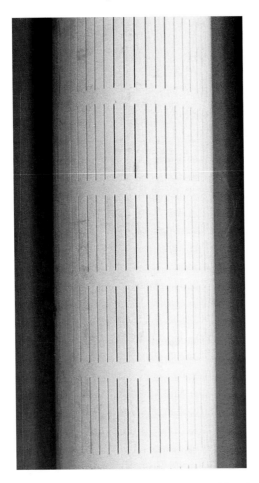

Figure 9.17. Casing has to be inserted in a borehole after or during drilling to prevent collapse; part of the casing has to be in the form of *screening*, in order to allow water to enter. Here the screening takes the form of a plastic tube with fine slits, but it can also be of metal mesh.

Above the seal, a *filter pack*, usually of medium-to-coarse quartz sand (which is chemically inert) is introduced via a pipe from the surface. The pack prevents fine particles of the natural wall of the borehole from entering the well via the screen slots, while also maintaining good hydraulic conductivity between the borehole wall and the screen. Above the pack is introduced a second seal of bentonite, thereby isolating the screen, hydraulically, at that precise depth.

When just a measurement of water level is required from a borehole, dedicated *piezometers* can be used. These are tubes of small diameter (2 to 5 cm) with a conical metal tip, behind which is a short tube of porous material, with a pore diameter of around 60 microns, acting as a screen. This is attached to a

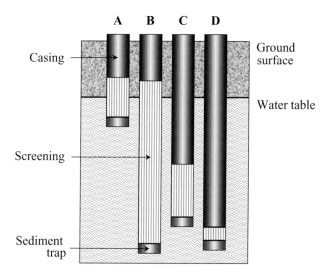

Figure 9.18. Four examples of how casing and screening may be installed are shown here. In A the screen extends across the full range of water-table changes. B covers the full saturated layer, C monitors a specific zone, while D is for a specific point.

tube that can be extended to any length required, giving access from the surface. These are much smaller, and so cheaper, than well casing and screening; they also require smaller-diameter, and so cheaper, boreholes. In complex situations, nests of piezometers are installed at different depths to detect vertical hydraulic gradients. Screens, or piezometers, may be installed at several depths in the same borehole to obtain a profile of water quality and piezometric head. Some piezometers are designed to be hammered into the ground (see 'driven wells', in the subsection on percussion drilling), which saves the cost of drilling a borehole, although this is only practicable under some circumstances such as near the surface or at the bottom of an existing borehole.

Costs

A drill rig with the necessary minimum of two operators will cost upwards of £300 a day depending on the type of rig. It can take from one hour to several months to drill a borehole, depending on circumstances. Boreholes are not, therefore, usually cheap and may well cost as much to drill as the cost of the instruments that are subsequently used in them.

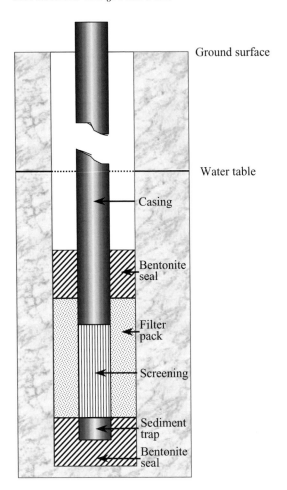

Figure 9.19. To ensure that only groundwater at the required depth is obtained for monitoring, the screen can be centred in the borehole and sealed above and below the required level with bentonite, with a chemically inert filter pack in between, to prevent particles from the borehole wall entering the screen.

References

Anderson, A. B. C. (1943) Method of determining soil moisture content based on the variations of the electrical capacitance of soil at low frequency with moisture content. *Soil Sci.*, **56**, 29–41.

Bell, J. P., Dean, T. J. & Hodnett, M. G. (1987) Soil moisture measurement by an improved capacitance technique, Part II. Field techniques, evaluation and calibration. *J. Hydrol.*, **93**, 79–90.

Bell, J. P. (1976) Neutron probe practice. Institute of Hydrology Report No. 19, p. 63.

Bell, J. P. (1987) Neutron probe practice. Institute of Hydrology Report **19** (third edition).

Bell, J. P. (1996) Determination of soil water content by the thermogravimetric method. Private communication.

Bloodworth, M. & Page, J. (1957) Use of thermistors for the measurement of soil moisture and temperature. *Proc. Soil Sci. Soc. Am.*, **21**, 1056–8.

Chadwick, D. G. (1973) Integrated measurement of soil moisture by use of radio waves. Utah State University, Logan, College of Engineering Report PRWG 103–1.

Clayton, C. R., Simons, N. E. & Matthews, M. C. (1982) *Site Investigation*. Granada Publishing, ISBN 0 246 11641 2.

Dasberg, S. & Dalton, F. D. (1985) Time domain reflectrometry field measurements of soil water content and electrical conductivity. *J. Soil Sci. Soc. Am.*, **49**, 293–7.

Davis, J. L. & Chudobiak, W. J. (1975) *In situ* meter for measuring relative permittivity of soils. Geological Survey of Canada. Paper 75–1, Part A, Project 630049.

Dean, T. J., Bell, J. P. & Baty, A. J. B. (1987) Soil moisture measurement by an improved capacitance technique, Part I. Sensor design and performance. *J. Hydrol.*, **93**, 67–78.

Dean, T. J. (1994) The IH capacitance probe for measurement of soil water content. Institute of Hydrology Report 125.

Debye, P. (1929) *Polar Molecules*. pp. 77–108. Dover, New York.

De Plater, C. V. (1955) Portable capacitance-type soil moisture meter. *Soil Sci.*, **80**, 391–5.

Dixon, A. (1998) Light percussive drilling. Private communication.

Evett, S. R. & Steiner, J. L. (1995) Precision of neutron scattering and capacitance type soil water content gauges from field calibration. *J. Soil Sci. Soc. Am.*, **59**, 961–8.

Gardner, C. M. K., Cooper, J. D. & Dean, T. D. (1998) Soil water content measurement with a high frequency capacitance sensor. *J. Agric. Eng. Res.*, in press.

Gaskin, G. J. & Miller, J. D. (1996) Measurement of soil water content using a simplified impedance measuring technique. *J. Agr. Eng. Res.*, **63**, 153–60.

Gilman, K. (1977) Movement of heat in soils. Institute of Hydrology Report 44.

Hasted, J. B. (1973) *Aqueous Dielectrics*, p. 302. Chapman & Hall.

Hoekstra, P & Delaney, A. (1974) Dielectric properties of soils at UHF and microwave frequencies. *J. Geophys. Res.*, **79**, 1699–708.

McPhun, M. (1979) Soil moisture meter. Report to the Department of Engineering Science, University of Warwick, UK.

Nadler, A., Dasberg, S. & Lapid, I. (1991) Time domain reflectrometry measurements of water content and electrical conductivity of layered soil columns. *J. Soil Sci. Am.*, **55**, 938–43.

Nielsen, D. M., ed. (1991) *Practical Handbook of Ground-water Monitoring*. Lewis Publishers, ISBN 0 87371 124 6.

Patten, H. E. (1909) *Heat Transference in Soils*. US Department of Agriculture Bur. Soils Bulletin. 59.

Robinson, M. & Dean, T. J. (1993) Measurement of near surface soil water content using a capacitance probe. *Hydrol. Processes*, **7**, 77–86.

Robinson, D. A., Bell, J. P. & Batchelor, C. H. (1994) Influence of iron minerals on the determination of soil water content using dielectric techniques. *J. Hydrol.*, **161**, 169–80.

Robinson, D. A. *et al.* (1998) The dielectric calibration of capacitance probes for soil hydrology using an oscillation frequency response model. *Hydrol. Earth System Sci.*, **2**, 111–20.

Shaw, B. & Baver, L. (1939) An electrothermal method for following moisture changes in the soil *in situ*. *Proc. Soil Sci. Soc. Am.*, **4**, 78–83.

Smith-Rose, R. L. (1933) The electrical properties of soils for alternating currents at radio frequencies. *Proc. Roy. Soc. London*, **140**, 359.

Strangeways, I. C. (1983) Interfacing soil moisture gypsum blocks with a modern data logging system using a simple, low-cost, DC method. *Soil Sci.*, **136**, 322–4.

Topp, G. C., Davis, J. L. & Annan, A. P. (1980) Electromagnetic determination of soil water content: measurement in coaxial transmission lines. *Water Res. Res.*, **16** (3), 574–82.

Whalley, J. R., Dean, T. J. & Izard, P. J. (1992) Evaluation of the capacitance technique as a method for dynamically measuring soil water content. *J. Agric. Eng. Res.*, **52**, 147–55.

Whalley, J. R. (1993) Considerations on the use of time-domain reflectrometry (TDR) for measuring soil water content. *J. Soil Sci.*, **44**, 1–9.

Wellings, S. R., Bell, J. P. & Raynot, R. J. (1985) The use of gypsum resistance blocks for measuring soil water potential in the field. Institute of Hydrology, Report 92, p. 26.

10

Water

Sensors for measurement of the quality and quantity of both surface water, including the oceans, and groundwater are similar in principle, and so it makes for greater clarity if this chapter is organised by sensor type rather than by application.

Measuring water level

Staff gauges

Graduated staff gauges are widely used for the manual measurement of rivers, lakes and sea level. They are usually installed vertically in the river bed or fixed to a weir (Fig. 10.1(*a*), (*b*)), bridge or harbour wall. Boards are made in one metre and two metre lengths and are about 15 cm wide, fixed one above the other to cover greater depths, and marked to span up to 12 metres, or more. Alternatively, several may be installed, each progressively higher up the bank of a river if there is no structure to which to fix them and the river is deep and wide. They are graduated in a variety of ways, some every centimetre, others every 10 or 20 centimetres – as in the case of some sea level gauges. Boards are also available for fixing at an angle of 45 or 30 degrees, laid flat on river banks, their markings being stretched to compensate. Most are graduated from bottom to top, but others are made with an inverted scale for situations where levels below a reference point are needed. Gauging boards have the advantage of cheapness and simplicity, although care is needed in reading them. Observers may send their readings by telephone or radio to a distant base, providing a simple form of telemetry.

(a)

Figure 10.1. (a) Staff gauges, as here at a trapezoidal flume on the Red River in Southern Papua New Guinea, are the commonest means of measuring water level, a manual reading being taken periodically. Some care is necessary not to mis-read the scale. (b) In the larger Iguacu River near Curitiba in Brazil, one staff gauge cannot cover the full range of level change and several must be stepped up the bank. This is a common procedure.

(b)

Peak gauges

While water level recorders, of the types to be described, give a detailed plot of the rise and fall of river level on each side of the peak flow, a device that records just the peak level of flow can be much simpler. A knowledge of the peak alone is valuable, but a fairly good estimate can also then be made of the shape of the hydrograph (from prior knowledge and experience) and thus of flow. This can be particularly useful when measuring events in ephemeral streams in the wadis of semi-arid areas. Here flow events are rare, but when they do occur the amount of water can be extreme and the speed of the event dramatic, often missed by observers who have insufficient time to get there. The amount of water flowing is important because it sinks into the ground, recharging the groundwater, which is often the only supply of water available in semi-arid countries.

The sticking-float method   A float, through which passes a vertical rod, rises as the river level rises, but as the level falls, the float is prevented from falling back down by a ratchet fixed alongside it. The float has a pointer attached to it which moves up a gauging board, the peak level being read after the flow has subsided; the float is then reset to the bottom.

The powder-in-a-tube method   An even simpler arrangement can be made in which a tube has powder, such as talcum or chalk dust, placed in its base (Fig. 10.2). As the water level rises, entering through a small hole at the base, the powder rises in suspension up the inside of the tube. As the level falls, a deposit of the powder is left sticking to a rod fixed within the tube, giving an indication of the highest level reached; the rod is then removed for reading, cleaning and replacement ready for the next event.

Stilling wells

Staff gauges stand unprotected in the water. But if a float is to be used to measure water level it must be protected from wind and waves. There are two main types of stilling well. In one a hut stands on the bank some distance from the river, protecting a well dug down to the lowest level the river will reach (Fig. 10.3). At its lowest level, a pipe connects the well to the river, the level of the water in the well slowly following the level changes in the river, with surface irregularities filtered out. This tube can become blocked, and care is needed to detect and prevent this. In an alternative design, often constructed of corrugated iron, the well stands directly in the river with a housing on top and walkway to it from the bank (Fig. 10.4).

Figure 10.2. Set in the bed of a wadi in northern Oman, the tube records the maximum level reached during the previous flow event (the flow can be short lived, and sudden in onset). Powder placed in the base of the tube rises to the flood level and sticks to a rod extending up the centre of the tube as the water subsides.

Figure 10.3. In the Red River of Papua New Guinea (see Fig. 10.1(*a*)), this stilling well is set back from the river, to which it is connected by an underground tube; this arrangement allows the hut to be accessed even at times of maximum water level.

Electrical contact gauges

Known as *contact meters* or *well dippers* (Fig. 10.5), an electronic probe is lowered by hand down the well (or borehole), detecting the water surface by completing an electrical contact through the water between the case and a point electrode. The probe is lowered on a reel of cable, marked in metres and centimetres, a unit at the surface emitting a light and/or a sound indication when the electrodes contact the water, the depth being read off the graduated cable against a datum point on the top of the borehole. It is possible to read depths to within ± 1 cm at 100 m depth, and dippers are made for use at up to 500 metres depth in boreholes. To support the weight of a long length of cable in the deeper boreholes, the wire is of steel.

Figure 10.4. In the Aquidauana River in the Mato Grosso area of Brazil, this stilling well stands in the river. It is accessed by foot at normal flows, over the walkway, and by boat if necessary when the river is in flood. Since it houses automatic instruments, access is less essential.

Some dippers also contain a resistance thermometer, allowing a temperature profile to be measured, while others may also include a conductivity, pH or dissolved oxygen sensor (see later). It is also possible to include a *ground detector*, which indicates when the probe touches bottom, the difference between the two readings giving a measure of water depth as well as the distance of the water surface from the top. Their use is not limited to boreholes, since they are just as functional in a river, lake or the sea. In some situations, a stilling well may not be necessary.

Figure 10.5. A well dipper is a simple means of measuring water depth manually, in stilling wells and boreholes. The probe (right) is lowered from a reference point down to the water. Upon contact, it sends a signal up the graduated cable, the depth being read from the cable. The reel shown here holds 200 m of cable.

Sensing water level by float

A range of methods, mechanical and electrical, of sensing and recording the position and movement of a float has been developed over the last 150 years.

Paper chart level recorders   Together with staff gauges, instruments which measure water level by float, moving a pen across a paper chart, account for most of the world's past river data and they are still in wide use today. The various designs are distinguished by three main features, as follows.

The paper chart can either have the water level axis horizontal (Fig. 10.6) or vertical (Fig. 10.7). The chart can wrap once around a drum (Fig. 10.7) or take the form of a long chart wound from one reel to another (Fig. 10.6).

The third main difference concerns how the pen moves across the chart. In the simplest design, the full range of level change is represented by the width (or height) of the chart. To record different level ranges, different gear ratios or pulley ratios are used. A typical chart is 25 cm high and is usually marked in 2 mm divisions. If the gear-reduction ratio is 1:1, then every 2 mm movement of the float will move the pen 2 mm across the chart, covering a total level change of 0.25 m. With a reduction ratio of 20:1, the range is extended to 5 metres, but each 2 mm movement of the pen now represents 4 cm float move-

Figure 10.6. This level recorder's chart is in the form of a roll, allowing extended recording times. The level is recorded horizontally (the time vertically), the pen being moved by a shaft (visible at the top of the chart) with a double-spiral groove, allowing the pen to reverse direction at the edges of the paper. This allows a wide range of level changes to be recorded to high resolution. Although this method is frequently and most commonly used with float recorders, the one illustrated here is of the bubble type. The pressure of the gas (representing the depth of water) is converted into shaft rotation by a servo-motor controlled by a mechanical balance that senses the bubble pressure via a bellows.

ment and this may not be good enough for some purposes; this is why the reversing-pen mechanism was developed.

By introducing a double-spiral groove in the rod that moves the pen (Fig. 10.6), the pen reverses its direction when it reaches the limits of the chart, allowing an unlimited range of level change to be accommodated without sacrificing level discrimination. The penalty for this is a record that is not immediately understandable and so, to aid in its interpretation, another pen marks the chart whenever a reversal of pen direction occurs, indicating if the level is rising or falling (Fig. 10.8). A reversal mechanism requires a larger float to overcome its extra friction. With a reversal mechanism, the range can still be varied through a selection of gears.

Accuracy starts at about $\pm 2\,mm$ for the small-range instrument, falling as the gear ratio (level range) increases. A well-maintained mechanism will perform better than a neglected instrument since friction will increase if it is not kept clean. (As with all pen recorders, the friction of the pen on the chart and the friction of the mechanical linkages has to be overcome.) The zero reference

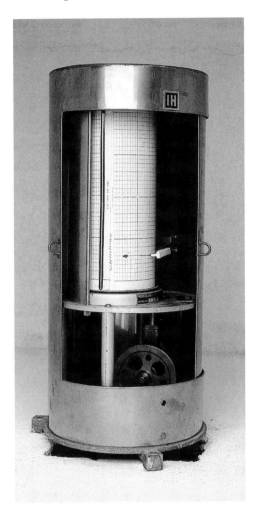

Figure 10.7. In this much simpler instrument, water level is recorded on the vertical axis (and time on the horizontal axis), the chart being a single sheet wrapped once around the rotating drum. The float moves the pen up and down the chart, the height of the paper (by choice of gear ratio) representing the maximum level change expected. When this is large, the resolution is low.

needs careful specification, but it is often arbitrary, such as a mark on a pipe, making it difficult to relate to levels at other points along a river. Height above mean sea level is preferred.

Time resolution depends on the speed of the chart. In the case of a drum, one revolution can be adjusted (again by changing gear ratios) to represent a time ranging from one day to one month, or longer, while for a reel-to-reel chart, from 2 to 60 mm per hour is selectable. To ensure appropriate timing accuracy, a third pen may be included to give a time-mark on the chart, triggered by the

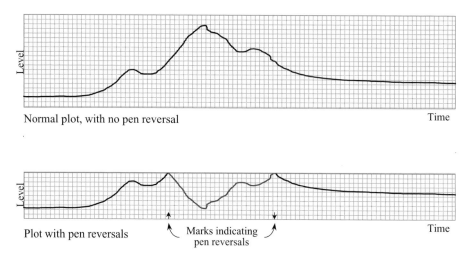

Normal plot, with no pen reversal   Time

Plot with pen reversals   Marks indicating   Time
                                pen reversals

Figure 10.8. By reversing the direction of movement of the pen when it reaches the edges of the chart, a wide range of level changes can be recorded to high discrimination on a chart of practical dimensions. The mechanism that achieves this is a double spiral cut in the rod that moves the pen (visible in Fig. 10.6).

clock mechanism for, even though the chart is time-marked, the paper can change dimensions slightly with humidity (this can also cause jamming if sprocket holes on the chart do not align precisely with sprocket wheels, for example through dampness). A refinement to the time-mark is to incorporate an independent electronic timer to trigger the pen.

A float turning a potentiometer   The simplest, and cheapest, way of converting float-movement into an electrical signal is to make it turn a potentiometer, which produces a resistance output proportional to water depth, this being converted into a voltage for logging. River level is one of the few variables that may justify a high-resolution sensor and logger because it is often necessary to measure water level changes of 1 or 2 mm over a range of 5 or 10 metres. Even 1 mm in 1 metre represents a resolution of 1 in 1000 (0.1%); 1 mm in 5 metres is 0.02%, representing 12-bit resolution (Chapter 11).

A potentiometer can take several forms. It can rotate just once, through about 358 degrees, or it can have a wiper that crosses the small gap (between the ends of the resistance track) and rotates continuously, or it can be of the multi-turn (10 to 20 turns) helically wound type. But a potentiometer in which the full range of level change is covered by just one rotation will only be able to achieve acceptable resolution when the change in level is small, say 1 metre or less.

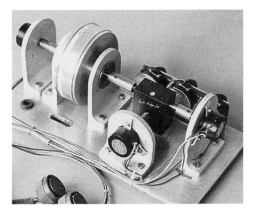

Figure 10.9. The simplest way of converting float movement into electrical form for logging is to make it turn a potentiometer. In the instrument shown, the pulley (left) is 20 cm in circumference, causing the first potentiometer (right) to turn once for a level change of 20 cm. At the far left, a second identical potentiometer is also directly turned, but the gap in its winding is 180 degrees out of phase with the first, to allow the sensor to operate over the full 360 degrees. A 20:1 gearbox (the black cube in line with the pulley) drives a third potentiometer (foreground) once for every 20 × 20 cm or 4 metres level change, so as to prevent errors if the level changes by more than 20 cm between recordings.

Where higher resolution is required, the same principle can be adopted as in the mechanical reversal of the pen of a chart recorder. In its electrical equivalent, the potentiometer rotates continuously, the wiper contact moving across the gap in the winding (Fig. 10.9). The float, in the example shown, turns a pulley with a circumference of 20 cm, which rotates a potentiometer once for each 200 mm change in level. As the potentiometer rotates its output rises until, as it approaches the gap in the winding, its reading is full scale. Thus, even at only 8-bit resolution, the 200 mm change can be measured to a resolution of better than 1 mm. After the gap is crossed, the reading returns to zero and starts rising again (if the water level is still rising), this going on indefinitely; thereby an infinite range is covered. To keep track of the overall (absolute) level, a second potentiometer, driven through a reduction gear box, rotates once over the full range of level change – or sufficiently slowly to avoid more than one rotation between loggings. A third potentiometer can be included with its gap 180 degrees out of phase with the first potentiometer, to avoid any error near the gaps in the windings. By using three logger channels, a high level-resolution is obtainable over a large level change. Such a technique works well even with loggers using only 8-bit analogue-to-digital converters.

However, if shaft rotation can be converted directly into digital form within the sensor, all the analogue errors disappear. This is the rôle of *shaft encoders*.

Floats turning a shaft encoder   Some shaft encoders are 'incremental', meaning that as the shaft rotates a switch opens and closes, the angle between
operations of the switch being proportional to a change in float level (of say
1 mm, or 1 cm). To take account of whether the float is moving up or down, two
switches are needed, 180 degrees out of phase. Which switch operates before
the other indicates the direction of rotation, the sensor electronics making the
necessary computation and giving an output of absolute level in straight
binary, binary-coded decimal (BCD) or the Gray code. (These codes, and their
strengths and weaknesses, are described in the section on analogue-to-digital
conversion in Chapter 12.) Because such sensors measure just changes in level,
not absolute level, they need to be programmed at installation with information as to the current water level, thereafter keeping track of the changes
automatically, unless the power is removed, when they must be reprogram-
med. The switches can be mechanical, but most modern encoders sense rotation magnetically or optically, these methods having less mechanical resistance
to float movement and a longer life. An incremental encoder, with electronics
to keep count of the absolute level, can cover as wide a range of level change as
required, to high discrimination, the shaft simply rotating many times.

Some shaft encoders, however, are not incremental but use a disc encoded
directly in the Gray code. In such designs the level changes which can be
resolved for a 360 degree rotation of the encoder shaft depend on the number
of (concentric) tracks on the optical disc. If the number is large, say 16, a wide
level range can be covered in just one revolution of the shaft, to high resolution.
In such a case, a reduction gearbox may be necessary between the float-pulley
and the shaft-encoder, so that only one revolution of the encoder results from
the full range of level change. This is of no matter, except possibly for backlash
in the gears. It is also possible to use a Gray-coded sensor, in a way similar to
the multi-turn analogue potentiometer level sensor, by allowing it to rotate
many times over the full range of level change, but this then loses track of the
absolute level.

Tone transmitters   Little used today, but useful in the past, and still needed in
some situations is the float-operated sensor that generates a series of tones for
transmission by radio or telephone, decoded by ear by an operator in a distant
office. In 1960s and 70s models, the float turned cylinders that were mechanically encoded with 0 to 9, one cylinder driving the next via a 10 to 1 gear
reduction. Typically there would be four cylinders, representing centimetres,
decimetres, metres and tens of metres. A microswitch, moved by a motor,
sensed each coded cylinder in turn, generating a string of four groups of pulses,

converted into voice-frequency tones for transmission. For example, a depth of 13.65 m would be represented by bursts of 1 tone, 3 tones, 6 tones and 5 tones, with spaces between each group.

More recently it has become possible to convert the BCD, or similar, codes generated by absolute shaft encoders into tones, thereby avoiding the complex mechanics of the early designs. Simulated voice announcements can also now be generated for transmission. In situations where just a few levels are to be measured and where there is no need for computer-based systems, this kind of instrument is an effective and simple way of using modern technology.

Combined methods   It is quite common to see two of the above techniques combined in one instrument, typically a shaft encoder or potentiometer being added to a paper chart recorder, thereby producing an on-site recording as well as being able to log or telemeter the measurements. Sometimes a shaft encoder and a pressure sensor are operated at the same site, one backing up the other in the event that one sensor fails, or the river freezes.

Sensing water depth by pressure

The bubble method   Nitrogen from a cylinder (or air compressed from a pump) is allowed to bubble slowly out of the end of a tube, the end being fixed in the river at the lowest flow level or suspended in a borehole below the lowest water level. The pressure of gas that causes gentle bubbling is equal to the pressure of the water above the outlet of the tube. By measuring the pressure, the depth of water can be calculated. The nitrogen gas must first be reduced in pressure from the high pressure in the cylinder. An adjustable valve then controls its speed of flow, the rate of flow being checked visually by bubbling the gas through a small water container on the instrument before it is piped to the vent point under the water. There are a number of methods of sensing the pressure and they fall into two groups – electromechanical balances and electronic diaphragm sensors.

Figure 10.6 shows an electromechanical balance in which the gas pressure, representing the river depth, operates a bellows that applies pressure to one side of a balance arm, while on the other side a weight is moved along the arm (by servo-motor) until balance is achieved. Balance is detected by an electrical contact on the arm, which stops the servo-motor. Movement of the servo-mechanism is transferred, via gears, to a pen which records a trace on a paper chart (with a reversal mechanism to enlarge the scale). A damping device, with a time constant of about 1 minute, is included in the tube to the bellows, to prevent waves on the water from causing the gas pressure to fluctuate rapidly. The accuracy claimed by manufacturers ranges from ±0.07% to ±0.1% of full range. These pressure sensors are not, therefore, as good as the better float-operated instruments, although when measuring level changes of a few metres the resolution is generally adequate, and even over large ranges this sort of accuracy is often sufficient. In a similar design, a mercury manometer is used in place of the bellows. A self-cleaning facility may be included that periodically applies a high gas pressure to the tube to blow out sediments.

By electronic pressure sensor

An alternative, and simpler, way of measuring water pressure is offered by electronic sensors, fixed in the river (or borehole or sea) at the lowest water level expected. Figure 10.10 illustrates the principle of this method, in which a diaphragm, in a tube, has the water pressure applied to one side, the other side being vented to atmospheric pressure (through its cable to the water surface). If the pressures on both sides of the diaphragm are the same (when the water level is zero), the diaphragm is flat and unstressed. As the depth increases, the diaphragm flexes uner the increasing pressure, the deformation being measured by straingauges fixed to the surface of the diaphragm – generally in the form of a bridge circuit. Today, the diaphragm is often of silicon, the straingauges being etched directly onto the surface using microelectronic techniques. Some sensors now use ceramic as the diaphragm material, which has better elastic properties than metal and a minimum of hysteresis, the deflection being measured capacitively (as Fig. 10.10 illustrates). Figure 10.11 shows an actual sensor. To prevent damage due to shock waves in the water and to remove very rapid fluctuations in level, the nose cap may contain a damper to suppress pressure transients, and filters to prevent sediment getting into the diaphragm cavity.

An advantage of using either electronic pressure sensors or bubble methods is that a stilling well is not essential and they can also be used in boreholes of small diameter and of any depth. Electronic pressure sensors, however, have the advantage over bubble-type systems of much greater simplicity. Some instruments, however, combine the two techniques, the gas pressure of the bubbler system being measured by an electronic sensor instead of mechanically. The advantages of this hybrid design are that a pressure sensor is simpler than a mechanical balance and that the electronic sensor is housed safely above water in the recorder housing, in the dry rather than in the river. Being close to the logger or chart recorder, long (expensive) cables are also avoided. The analogue voltage output from diaphragm sensors can, after suitable processing, drive an electrical chart recorder, display the level visually,

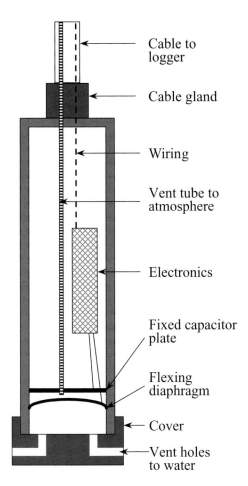

Figure 10.10. The pressure of the water overhead, and thus its depth, can be measured by sensing the deflection of a diaphragm, one side being exposed to the water pressure, the other to atmospheric pressure. In this schematic figure, the movement of the diaphragm is sensed by measuring the capacitance between it and a fixed plate, the associated electronics being housed close by in the body of the sensor. Some sensors use different methods of measuring the deflection.

be recorded on a logger or be telemetered to a distant base – or a combination of these possibilities.

There are, however, disadvantages to the diaphragm sensor. At best an accuracy of ±0.05%, and more typically ±0.1%, of full range is achievable. (This makes them very similar in performance to mechanical bubbler instruments.) If additional errors are introduced in the sensor electronics, and in the analogue-to-digital conversion process, a more realistic accuracy might be closer to ±0.3%. Thus, while the resolution of a flexing diaphragm is infinite, if

Figure 10.11. With its protective cap (left) removed, the diaphragm of this pressure sensor is visible (right); this flexes under the changing weight of the water above it, giving an electrical output proportional to depth. The cap contains vent holes to allow the water pressure to be sensed while protecting the diaphragm from damage.

the level change to be measured is, say, 10 metres, the best accuracy achievable is about ±5 mm, and a more realistic performance is ±20 mm. Pressure sensors are thus not as good as the best float-operated instruments, and, although it is less true of today's pressure sensors, there might be a long-term calibration drift. Nevertheless, because of their convenience and relative cheapness they are becoming widely used.

Sensing water level by ultrasonics

Not widely used for sensing water level in environmental applications are ultrasonic sounding systems (compare the method for measuring snow depth in Chapter 8), which sense the time of flight of an ultrasonic pulse from a transmitter to the surface of the water and back to a receiver, the sensor either being above the water looking down or in the water looking up. More details of this are given later in this chapter under ultrasonic flow gauging.

Measuring water current velocity

A brief history

The first river current meter was probably made by the Italian physician, Dr Santorio, around 1610, and consisted of a plate held vertically in the stream on a balance arm, the pressure of the water on it being measured by a movable weight (Frazier 1969) although the mathematical tools to convert the pressure into velocity had not yet been devised. Robert Hooke also designed a current meter, about 70 years later, for measuring the speed of ships through the sea, as well as the current of rivers. This was based on a propeller, reconstructed by Frazier (1969), that looked remarkably like a present-day current-meter propeller. Paolo Frisi, professor of mathematics at the University of Milan in the eighteenth century, described a paddle wheel for measuring surface velocity (Leliavsky 1951, 1965). It should be remembered that at this time even the origin of rivers was uncertain; Frisi did in fact believe that they derived their origin from the rain, although he was wrong in the belief, shared by others, that river currents changed in speed parabolically with depth (Herschy & Fairbridge 1998).

The Pitot tube (Fig. 5.7) was developed specifically to measure river velocity (De Pitot 1732). However, Pitot was unable to calculate correctly the velocity of the current from the height to which the water rose in his instrument, because the value for g had not yet been determined and so the correct equation $v = (2gh)^{1/2}$ could not be derived for the Pitot tube. Nevertheless the instrument was very useful in determining the relative velocities of currents at different depths, and proved the fallacy of the formerly held view that it changed parabolically with depth; exposing the error of this view was itself an important contribution made by Pitot.

Finally, Reinhard Woltman, a German engineer, designed a current meter in 1790 in which two angled plates on a cross-arm turned a shaft connected to a revolution counter (Frazier 1969). According to Biswas (1970), for many years after Woltman's death each improved version was still called a Woltman current meter out of courtesy. It was, in effect, a crude propeller.

Mechanical current meters

A modern mechanical current meter (Fig. 10.12), senses flow through the water impinging on a propeller, or impeller – a better word might be 'helicoid' since it is helix-shaped. The helicoid rotates a shaft on which there is a rotation detector – usually a magnet operating a reed switch, although optical methods are also now used. The pulses produced by the meter are counted, either by an electromechanical counter (which is limited in the maximum speed to which it can respond – this is around 10 counts per second) or by an electronic counter (which has no speed limit). In a design that is an alternative to pulsed outputs, but is again comparable with a wind speed sensor, the propeller rotation is converted into a varying DC voltage by a small generator. This allows a continuous strip-chart record to be made of the flow. Although current meters can be permanently installed in a river, they are normally used as portable instruments operated manually – hand-held on graduated *wading rods* while

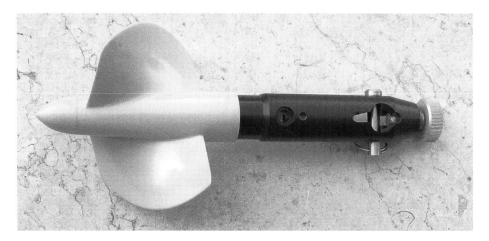

Figure 10.12. Although made in a range of sizes, all mechanical current meters sense the movement of water by means of a propeller in the shape of a helicoid, generally with two or three blades, the pitch varying to suit different velocities. A magnetic reed or an optical switch produces pulse signals as the blades rotate.

the operator walks across a shallow river, or for deeper waters suspended from a jib on a bridge or boat or from a cableway.

Depending on whether light or strong currents are to be measured, the size of the instrument can be varied, the smaller measuring currents up to about 5 m s^{-1}, the larger up to 10 m s^{-1}, typically with accuracies of about $\pm 2\%$ of full scale. The helicoid propellers range in diameter from 5 to 12 cm and are made of metal or plastic. In addition they vary in pitch, expressed in metres. Pitch is not a characteristic that can be measured physically on the helicoid itself; a pitch of, say, 0.1 m expresses the fact that one revolution will be produced by the passage of 10 cm of water. One revolution per second would, in this example, mean that the current velocity is 0.1 m s^{-1}. Pitches vary from 0.1 to 1.0 m and have to be calibrated in currents of known velocity. Pitch is a characteristic of the shape of the helicoid and once a particular shape has been tested it is assumed that all helicoids of that design will behave in a similar way. It is usually possible to interchange impellers on the same instrument to measure currents of different speeds. There is a minimum speed below which the instrument will not respond, typically around 0.03 to 0.05 m s^{-1}, limited by friction at the bearings. Also, as with helicoid wind sensors, the response of a current meter is dependent on the angle of attack of the current and can be assumed to have a cosine response over a small off-centre alignment. When suspended from a jib or cableway, meters are provided with a tail that points them into the current. In stronger currents it is also necessary to add weight to the meter so as to keep its suspension cable reasonably vertical. The *sinker weights*, which vary from 5 to 10 kg, fit just below the meter and are streamlined.

Ultrasonic current meters

It is possible to measure the flow of water in a river by the Doppler shift in frequency of a continuous sound wave emitted upstream, reflections from bubbles and small particles in the water providing a returned signal of shifted frequency. Hand-held instruments using this principle are now available, in the style of a propeller meter, although they are less common than the electromagnetic meters described next. Because the method operates by emitting a cone of sound into the stream of water, it senses the velocity of the cone of water, not, like the propeller-type, just the water with which the blades come into contact. Typically an ultrasonic meter will sense water velocity up to about one metre upstream within a cone of about 15 to 30 degrees. It is not, therefore, so position specific as a propeller. Because flow in rivers contains eddies, an instrument of this type is designed to take many readings, up to 1000, and to determine the mean velocity from a potentially quite wide velocity spectrum. In contrast, propeller sensors have a natural averaging response, being driven by the sum total of the eddies impinging on the blades, rapid changes in rotation being prevented by the inertia of the mechanism.

The accuracy quoted by manufacturers is between $\pm 2\%$ and $\pm 5\%$ of the measured velocity (not of full scale) and the range covered is typically from 0.1 to $5\,\mathrm{m\,s^{-1}}$. The resolution is around $1\,\mathrm{mm\,s^{-1}}$.

Impeller and Doppler current meters are not intended for permanent installation in a river. Where continuous current metering is useful, however, a larger design of Doppler instrument is now available for permanent installation on the river bed. As there are no moving parts in ultrasonic systems and thus no mechanical wear-and-tear, they are better suited to continuous operation than their mechanical counterparts.

Electromagnetic current meters

Michael Faraday showed that a conductor moving in a magnetic field has a voltage induced across it proportional to the length and speed of the conductor and to the strength of the field, such that $E = BVl$, where E is the induced voltage (emf), B is the strength of the magnetic field in tesla (symbol T; 1 tesla = 1 weber $\mathrm{m^{-2}}$), V is the average velocity of the conductor in metres per second and l is the length of the conductor in metres. By making the conductor the river water, the voltage induced in it through its motion enables its velocity to be measured. Several hand-held current meters are now based on this concept.

Figure 10.13 shows how water 'filaments' cut through the magnetic lines of force as the water moves, and so generate a voltage across its length. In electromagnetic flowmeters, the magnetic field is generated by a coil powered

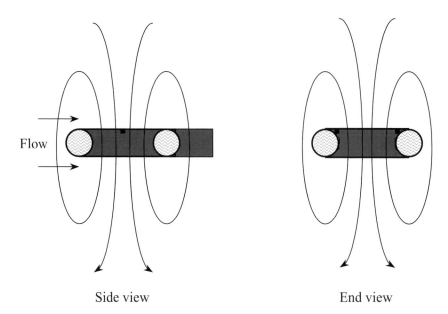

Side view   End view

Figure 10.13. A smooth, round-ended, epoxy resin body, about 2 cm thick, 5 cm wide and 20 cm long, is held so as to face into the water flow. The rounded end contains a coil (light shading), in the plane of the flow, that induces a vertical AC magnetic field through which the water flows, inducing a voltage across the conducting 'filaments' of water; this is picked up by the two titanium sensing electrodes (small black squares) and amplified, giving a measure of current velocity.

by a square-wave current with a frequency of about 130 Hz. By using an AC signal, and thus a varying magnetic field (rather than a DC current and a fixed field), it is easier to amplify the resulting AC signal generated by the movement of the water through the field. Small DC signals are more difficult to handle than are AC, suffering more from zero drift and random noise.

Readings are taken several times a second and averaged over periods from 5 to 100 seconds. At the upper end of the averaging periods, the accuracy is in the region of $\pm 0.5\%$ of the reading (not of full scale). The range usually covered is up to $\pm 5\,\mathrm{m\ s^{-1}}$, although instruments measuring up to $\pm 10\,\mathrm{m\ s^{-1}}$ are available (forward and backward directions are measurable because if the direction of flow reverses so too does the polarity of the induced voltage). Although the more usual instrument has just one pair of sensing electrodes, by introducing a second pair at right angles to the first it is possible to separate the velocity into two vectors at 90 degrees to each other and thus to measure both the velocity and the direction of the current. Sensors to measure current in three dimensions are also manufactured.

As with ultrasonic meters, the volume of water being measured is less clearly defined than in the case of a propeller sensor, since the conductor (the water), unlike a piece of copper wire, is not clearly bounded but extends out beyond the electrodes into the surrounding water; this is in part dependent on the conductivity of the water. So it is necessary to calibrate sensors of this type in the laboratory in order to arrive at a relationship between current velocity and induced voltage; once this is done for a particular design, however, it is the same for all instruments of that design, although conductivity-dependent.

Because the voltage generated by the movement of the water is measured by an amplifier with a high input impedance, no electrical load is placed on the source of voltage so very little current is drawn from it and, as a result, it is the open-circuit voltage that is measured. In consequence, the conductivity of the water (which can be visualised as a resistor in series with the source) has no effect on the readings, although below a conductivity of about $5 \ \mu S \ cm^{-1}$ its functioning may be impaired. As the current velocity approaches zero, the generated voltage becomes exceedingly small and at velocities below a few millimetres a second the exact zero point becomes uncertain.

Measuring river discharge

Apart from knowing the depth of water in a river, which is an important variable in itself (for example, in flood warning), the quantity of water flowing, the volumetric flow or discharge, is equally important. Discharge is expressed in cubic metres per second ($m^3 \ s^{-1}$), abbreviated to 'cumecs', although at very low flows in small streams litres per second might be more convenient.

Stage-discharge rating curves

If the cross-sectional profile of a river is known, a measurement of the depth of water (the *stage*) at any moment will give the area of water flowing and, if the velocity is also known, the discharge can be calculated by multiplying the cross-sectional area of water by its mean velocity. In practice this is done by using a current meter to measure the velocity at several depths at each of a set of *verticals* across the width of the river (Fig. 10.14, on the left). In all but the smallest rivers, from 20 to 30 verticals are selected, the spacing between them generally being chosen so that they each represent a segment of about equal flow. No segment should contain more than about 10% of the overall flow. In rivers wider than 300 m, the verticals are usually spaced equidistantly.

The total discharge is equal to the mean velocity in the segment times the width of the segment times the depth of the water. The total discharge of the river is the sum of each segment's flow. From the measurement of discharge

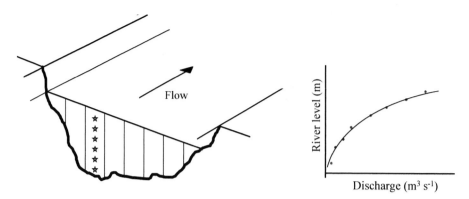

Figure 10.14. To produce a stage-discharge rating curve (right-hand figure), the river is divided into about 20 to 30 verticals (only eight are shown here), and in each segment the velocity is measured at several depths (star positions) with a current meter, producing an average discharge estimate for each segment. These are summed to give a total discharge for the river, this procedure being repeated for different levels of the river. From the readings, a plot of discharge against depth – a rating curve – is drawn.

and depth, a *rating curve* is drawn smoothly through the points obtained by current metering (Fig. 10.14, on the right), allowing discharge, thereafter, to be measured simply by measuring the depth of the river (Herschy 1995).

As explained in the previous section, depending on the size and depth of the river the current meter may be hand-held, the operator wading across the river, or it may be deployed from a bridge or a boat, the meter being lowered by hand or from a winch with a jib. Measurements can also be made by suspending the current meter from a permanent cableway installed across the river from bank to bank (Fig. 10.15).

If the profile of a river changes with time, the rating curve also changes. This is particularly the case after flooding, when the banks and bed may be suddenly and perhaps considerably altered. When this occurs, the field work has to be repeated and a new curve produced, which can be an expensive process involving a team of field workers at remote sites for days. Normally the relationship should be checked several times a year.

While it is fairly easy to obtain a rating curve for low to medium flows, it is often impossible to get to a river to make current meter measurements when flooding is occurring, and so it is usually necessary to extend the stage–velocity curve up to flood stage, beyond the level to which current metering has been possible. This procedure is based on a study of flow characteristics (Corbett *et al.* 1945), which includes allowances for the slope of the river, the presence of a contraction or a widening downstream, bends in the river, downstream tributaries, dams and plant growth. Even so, at very high flows, when the river overflows the banks, it may be difficult to establish what the flow is.

Figure 10.15. If wading in a river is not possible and there is no boat available or bridge conveniently in place, current metering must be done by suspending a current meter (Fig. 10.12) from a cableway. The meter is suspended from the fixed upper cable on a crane trolley, the two tow cables drawing the meter across the river, where it is stopped at several points and lowered to several depths. When suspended rather than being hand-held on a rod, the meter is fixed to a sinker with a stabilising tailpiece to keep it at the required depth and facing into the current.

An alternative to actual current metering is to make measurements after the flood event. The most commonly used technique, the *slope-area* method (Dalrymple & Benson 1967, Riggs 1976), involves surveying the slope of the debris left by the flood (the *silt line*) and obtaining an estimate of the *bed-roughness coefficient*, through a knowledge of the bed material and the size distribution of the bedload. To compute river discharge, the *Manning formula* is often used (Barnes & Davidian 1978, Limerinos 1970). Here the discharge $Q = AR_h^{2/3}S^{1/2}n^{-1}$, where A is the cross-sectional area, R_h is the hydraulic radius (the cross-sectional area of the river divided by its wetted perimeter), n is the bed-roughness coefficient and S is the friction slope (the change of water-surface level over a distance). As this method does not involve the use of instruments, it will not be expanded on further.

Weirs and flumes

To overcome the problem of changing profiles, it is sometimes possible to route the river through an artificial structure, the shape of which does not change with time. Knowledge of the precise shape also makes it possible to calculate a theoretical rating curve (Horton 1907).

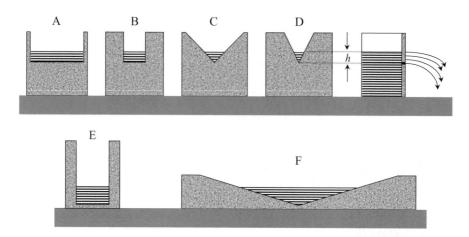

Figure 10.16. The height of water *h* behind a weir (A to D) gives an accurate measure of the outflow velocity and, since the area of the notch is known, the discharge can be calculated. A is a weir extending across the full width of the rectangular approach channel while B, C and D are contracted weirs with different cross-sectional shapes to suit different flow rates (see text). For correct operation there must be atmospheric pressure under the sheet of water (the nappe) flowing over the weir. E and F are flumes, rectangular and trapezoidal respectively, which control the shape of the channel, stabilising the rating curve. The advantage of the trapezoidal design is that the range of measurable flows is wide.

Weirs   A weir is a dam constructed across a stream or river, controlling the depth rather than the width of the channel; the water flows either through a hole cut in a plate across the channel or it may flow across the full width of the channel, as Fig. 10.16 illustrates. The shape of the outflow hole is usually either rectangular or V-shaped, the angle of the triangular V-notch varying from 90° downwards. The most usual alternative angles are those known as 'half-ninety' (53° 8′) and 'quarter-ninety' (28° 4′), the half and quarter referring not to the angle but to the area of the notch relative to one of 90° – and thus to the relative flows they can deal with. V-notch weirs allow small flows to be more accurately measured than rectangular-notch weirs. Although provision is usually made to prevent the build-up of debris, any blockage of the outflow will affect the readings and the narrower-angled notches are more prone to this. The measurement of flow is achieved by measuring the depth of water behind the weir plate, the velocity of the outflow being theoretically equal to $(2gh)^{1/2}$, where *g* is the acceleration due to gravity and *h* the measured depth of water above the crest of the weir. As the area of the notch through which the water flows is known precisely, the flow can be calculated from a measurement of the head only (Ackers *et al.* 1978, Herschy 1995).

The most commonly used weirs are the broad-crested (square-edged) weir,

Figure 10.17. This small flume at the Institute of Hydrology experimental catchment in Plynlimon, central Wales, is designed to cope with the fast-flowing steep streams encountered in upland areas, with provision to slow the flow temporarily and to deal with bedload movement.

the rectangular-profile weir and the triangular-profile weir, these being design-ed individually to suit the channel and flow conditions. Although weirs are more suitable for small streams than large rivers, several designs are available for larger channels and natural rivers; they are usually constructed in concrete.

Flumes   A flume is a channel through which a river passes. A river bed is a natural flume, but the term is usually applied to an artificial channel. The cross-section of a flume is usually either rectangular or trapezoidal (Fig. 10.16) and as Figs. 10.17 and 10.18 show, they vary in size and circumstance consider-ably.

Even though laboratory rating curves have been developed for flumes and weirs, it is usual to confirm their actual performance by field tests. These tests might use current meters or other methods such as dilution gauging, which we now discuss.

Figure 10.18. In contrast to the previous figure, this trapezoidal flume in the path of a wadi in Oman has to measure over a range from the low flow seen in the figure up to large flood events. The instrument stilling well is visible on the end of the wall of the flume (to the left of the two men).

Dilution gauging

Where river flow is very turbulent, such as in steep streams in mountains, the use of a current meter may be difficult or impossible and may give unreliable estimates. In such situations, *dilution gauging* may be more suitable (Water Research Association 1970, Gilman 1977).

A chemical tracer, in the form of a solution of a salt, for example an iodide or a lithium salt, of known concentration C_1, is injected into the river at a known and constant rate q. After a time, sufficiently far downstream to allow thorough mixing, the tracer concentration will reach a plateau C_2, which can be measured. Knowing these values allows the quantity of water flowing in the river, Q, to be calculated from $Q = qC_1/C_2$. If the tracer occurs naturally in the river, its natural concentration C_o must be subtracted from the measured concentration C_2 of the samples. This may only need to be done once, although it is wiser to check it at different flows by sampling upstream of the injection site. To ensure a constant injection rate, a 'Mariotte' bottle is used; this gives a constant head of liquid as the level in the bottle falls (Fig. 10.19). The flow rate is checked by weighing a timed release. To ensure that the plateau of concentration has been reached, samples need to be taken at intervals over a sufficiently long period (a few hours), this either being done manually or by automatically timed samplers. The full process of injection and

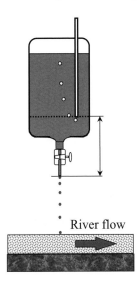

River flow

Figure 10.19. To ensure a constant injection rate of the tracer solution into the river for dilution gauging, a Mariotte bottle can be used: the constant pressure head, shown by the double-arrow pointer, is achieved by having a vent tube to the atmosphere as shown.

sampling can be completely automated so that when the river reaches a preset level the injection and sampling can be started. At the same time, the level of the river must also be measured so that a rating curve can be derived. When the rating of the river is completed, the dilution gauging can be stopped. An automatic dilution meter, which uses common salt as the tracer and which can measure up to $4 \, \mathrm{m}^3 \, \mathrm{s}^{-1}$, is now available.

Ultrasonic flow gauging

As well as being the basis of hand-held current meters, ultrasonics can also be used in permanent installations in large rivers to give a continuous measurement of volumetric discharge. But instead of sensing the velocity at a single point, it is measured across the full width of the river, in this case not by Doppler shift but by the flight time of an ultrasonic pulse. In its simplest configuration, two piezoelectric ultrasonic transmitter–receivers are fixed in the river on opposite banks, one upstream of the other (Fig. 10.20(*a*)). Both simultaneously emit an ultrasonic pulse and then immediately switch functions to receive the pulse arriving from the transducer on the opposite bank, the pulses propagating diagonally across the river in opposite directions. The pulse travelling downstream, with the current, arrives more quickly than that going upstream. The flight times (t_d and t_u) are measured.

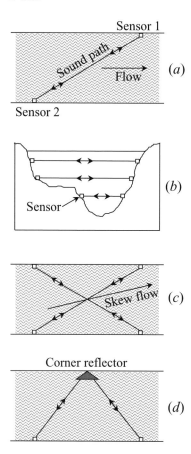

Figure 10.20. Ultrasonic river flow gauging. (*a*) Plan view: the simplest arrangement uses one pair of sensors (small white rectangles) at one depth on opposite banks. (*b*) Cross-section: in deeper rivers several levels may have to be measured, with pairs of sensors at each depth. (*c*) Plan view: for better accuracy, particularly where the flow of the river might be skew, on a curved reach, a crossed-path array is used. If having sensors on two banks is not practical, a reflector can be used (*d*).

If the length of the sound path is L metres, then $t_u = L/c - v$ and $t_d = L/c + v$, where v is the velocity component of the current in the direction of the sound path and c is the speed of sound (at the temperature and water density at the time). Taking the difference of the flight times,

$$t_u - t_d = L/c - v - L/c + v$$

This reduces to

$$v = (L/2)(1/t_d - 1/t_u)$$

which is independent of c. Furthermore, $V = v/\cos \alpha$, where V is the velocity of the stream parallel to the banks.

Taking the sum of the times of flight $t_u + t_d$ gives

$$c = (L/2)(1/t_d + 1/t_u)$$

which is the actual speed of sound at that moment under the particular conditions. This is useful if river depth is also measured by an (upward-looking) ultrasonic sensor on the river bed – for, in addition to measuring the current, the river depth also has to be measured so as to convert the velocity data into volumetric discharge. While it is possible to use an alternative method to measure river level, it can be convenient to use an ultrasonic method since suitable circuits are already to hand; the availability of c from the current measurements is then useful for the adjustment of the depth reading.

If the river is deep, it is necessary to instal pairs of ultrasonic transducers at several depths (Fig. 10.20(*b*)), each path being measured in succession (Davis 1979). Often the exact angle of flow may not be known, or it may be known to be skew (not parallel to the banks). To deal with this problem, a crossed-path array is used (Fig. 10.20(*c*)), the two readings helping to cancel out any non-parallel flow.

It can sometimes be inconvenient to have transducers on both sides of the river. To get around this difficulty, a reflector can be installed on the far bank, instead of a transducer (Fig. 10.20(*d*)), the sound echoing off it to the second transducer, further along the same bank. A secondary advantage of using a reflector is that it produces an effect similar to a crossed-path array and so helps overcome any skew in the flow. A corner reflector is used because it is not so critical to alignment with the sound beam as a flat plate is. (The reflector has to be kept clean in order to operate reliably.) The angle of the sound beam is a cone of about 10°, although there is spread up to $\pm 25°$ with some additional side-lobes in the polar diagram (Herschy 1979a).

The sound frequency used ranges from 30 kHz to 1 MHz, depending on path length, the lower frequencies being less affected by sediment suspended in the water and so more suited to wider rivers and those with heavy sediment loads. The radiated audio power ranges from 500 watts for rivers under 50 metres width to 1 kW for anything wider, although these are pulsed, not continuous powers. Nevertheless this makes the technique higher in power consumption than most instruments so far considered and mains power is almost an essential, although isolated examples use large 12 volt batteries, perhaps charged from the mains or by wind power. It is a heavy-weight system in all senses, including cost; this starts at £15 000, rising to well over £50 000. Because cost does not increase with river width, it is best used for wide rivers (rather than the electromagnetic method described below in which cost is very much width-dependent).

It is difficult to be precise about accuracy because it depends on the width and depth of the river, on whether it is rapid or slow moving, turbulent or placid and on how many levels of acoustic path are used (Herschy 1979b). But manufacturers quote around 5% of full scale for the current velocity measurement. Because the method measures velocity and depth continuously (or very frequently) it will inevitably produce better results than the much cheaper and simpler indirect techniques that measure just the level and rely on empirical stage–discharge curves to convert level to volumetric flow. But this higher precision is obtained only at higher cost and so the technique is something of a rarity.

Nor is it suited to all situations (Walker 1979). The growth of plants can affect the passage of the sound, as can the passage of boats, the entrapment of air bubbles or variations in the density of the water owing to temperature gradients or salinity. Nor can it be used in shallow water or, for the same reason, near to the bed or the surface, because of reflections from them. An alternative, free of these problems, but only feasible for the smaller river, is the electromagnetic method.

Electromagnetic flow gauging

A brief history   The Faraday principle (Fig. 10.13) as applied to hand-held current meters (see earlier) can also be used to measure the velocity of the complete river, continuously and automatically. Indeed it was Faraday himself who noted that water flowing in the earth's magnetic field induces a voltage which can be detected by two electrodes (Faraday 1832). The induced voltage in each filament of water is directly proportional to the average velocity of the water as it moves through the vertical component of the earth's magnetic field.

This principle was used in 1953 to measure the flow through the Straits of Dover (Bowden 1956), using a telephone line to pick up the voltage induced on the French side of the Channel. It was also used two years later to measure the tidal flow in the River Humber (Cox 1956). However, the extremely low potentials that are generated ($5 \mu V$ per metre width of river) and the fact that the signal is DC prevent the method being used in small rivers, owing to the swamping effects of the much higher levels of interference that occur. These are caused by electrochemical and thermoelectric effects, by mains 'hum', by radio signals and by boats, as well as by noise originating in the amplifier electronics.

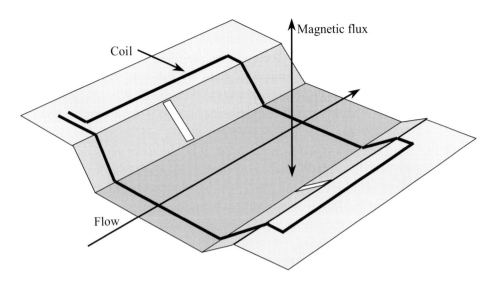

Figure 10.21. Similar in principle to a hand-held electromagnetic current meter, but of greatly increased scale, an electromagnetic river-flow gauging station generates an AC magnetic field spanning the whole river, the flow of water through it generating a voltage from bank to bank which is picked up by the two electrodes (the white rectangles), the amplitude of the signal being proportional to velocity. Knowing the cross-sectional area of the river, through a measure of its depth, the discharge can be computed. It is best suited to smaller streams.

A practical system   With the advance of electronics, in the 1970s the development began of a technique that uses artificially induced, much larger, and alternating magnetic fields (Fig. 10.21) to measure river flow electromagnetically (Herschy & Newman 1974). In principle it is identical to that of the electromagnetic hand-held current meter, simply scaled up. For the same reasons, of better zero stability and immunity to noise, the magnetic field is again induced by an AC signal.

However, the scale is hugely different: the coil has to be slightly larger than the width of the river and about as long as it is wide. It is usually laid on the bed of the river, although, if circumstances permit, it can be over the water. While the frequency of the square wave used in the small hand-held current meters is around 130 Hz, the frequency here has to be low enough to avoid any capacitive and inductive pick-up by the electrodes, but high enough to allow differentiation from noise. This turns out to be around 1 Hz. The coils that carry this square wave are from 50 to 300 turns of insulated cable of about 3.5 mm diameter, carrying a pulse of 3 to 5 amps at a voltage of 24–240 volts with a power of 5–1000 watts. Because the coil is an inductive load, the magnetic energy stored in it has to be dissipated at each reversal of pulse polarity. This is done using the conventional capacitor and diode circuits, which results in a back emf voltage spike at the moment of reversal. This is allowed to die away, and the magnetic field to stabilise, before the signal voltage (the induced voltage across the river) is sampled.

The signal is picked up by two stainless steel strips, laid on opposite banks. In this way the full depth of the river is measured, unlike in the ultrasonic method where discrete levels only are sensed. The construction is simpler if the river is channelled through a flume or vertically sided containment.

The electrical conductivity of the river bed can be quite high, thereby reducing the signal level by drawing current from the 'source' and so producing a voltage drop across the source's internal resistance. There are two ways to deal with this. Either the bed can be insulated from the river by laying a plastic sheet beneath the coil and under the signal pick-up strips, or the bed and river-water conductivities can both be measured and a correction made. The latter is less desirable than the use of an insulating sheet because if the bed conductivity is high the signal level may be reduced so much as to make its reliable detection from amongst the noise difficult.

Evaluated against several other methods of volumetric flow measurement, the electromagnetic method shows an accuracy of $\pm 2\%$ to $\pm 5\%$ of the actual flow. However, owing to the complexity of the equipment and the civil engineering requirements of installation, costs range from £20 000 upwards. As with the ultrasonic method, the electromagnetic technique is only used where the highest precision is required. Throughout the world, the velocity–area method using a current meter, combined with a stage–discharge curve, remains the simplest and by far the most used technique to measure volumetric flow, and where the rating curve is stable, it is probably as good as any for most practical purposes.

Measuring water quality

The variables

The term *water quality* includes the bacterial content of water, the presence of oil in water and a host of ions, in particular calcium, magnesium, ammonium, nitrite, fluoride, phosphate, sodium, chloride and potassium as well as trace metals such as aluminium, iron, copper and lead. But, in practice, there are just about six water quality variables that are of regular and universal concern, and it is these on which we will concentrate here. They are: water temperature, conductivity (which merges into salinity), pH, dissolved oxygen (DO), bio-chemical (or biological) oxygen demand (BOD) and suspended solids (merging into turbidity). The many other water quality variables are related more to public health, land-fill monitoring, industrial and waste water and potable water rather than to natural rivers and groundwater.

Method of measurement

The options   There are six principle ways of measuring the quality of natural waters.

- Portable instruments can be submerged in the water and readings taken manually on site.
- Samples can be collected manually and measured immediately *in situ* with portable instruments.
- A sample can be collected manually in a container and taken back to a laboratory for tests.
- Samples can be collected automatically for later laboratory analysis.
- Submerged instruments can be left unattended and their measurements logged or telemetered automatically.
- Water can be pumped from a river on a continuous or intermittent basis into a hut or small self-contained unit containing automatic sensing and logging systems.

An intercomparison of methods   Some variables cannot be measured in the laboratory; temperature, for example, must obviously be measured in real time, preferably directly in the water. But other variables such as pH and DO may also be affected by storage, particularly if it is long term in bottles, owing to interaction with the container, carbon dioxide in any air in the container and biological activity, although these can be reduced by the correct choice of materials and, at some expense, by cooling. Filling the container completely in order to exclude air also helps. Pumping can also affect the sample: heavy sediment may not be transported fully, the temperature will change and pressure changes due to pumping may affect dissolved gases such as oxygen.

   In some cases, the only practical method is to collect a sample. For such situations, special samplers are available that open at the required instant or take a sample at the appropriate depth. If the samples are measured quickly, the results will generally be acceptable. Automatic samplers, containing perhaps 24 bottles under vacuum, or with a peristaltic pump, are also used; a timer opens the bottles, or starts the pump, at preset intervals, allowing the water to be drawn in. Some designs allow samples to be collected when preset events occur, such as when a given river velocity is exceeded. Pumped systems may have to back-purge the tubes prior to taking a sample, so as to displace the water retained since taking the previous sample. But sample collecting will only suit those variables that do not deteriorate or change under storage.

The submersion of sensors directly in the river, borehole or sea avoids many of these problems, but if the sensors are left to run unattended, many of them quickly lose their calibration (within days) – much more quickly than any of the meteorological or water quantity sensors. If the sensors are deployed out of the river in a hut (or in a more compact container) with the water pumped to them, it is possible to clean and recalibrate them automatically, although this is very complex and expensive. Which method is used will depend on the variables being measured and on the nature and source of the water. For example, a submerged pH sensor may perform perfectly for months in the stable cool water of a deep borehole, away from light and algae growth, whereas it might drift off calibration in a few days in a tropical river.

If samples are returned to a laboratory for test, the same sensors can still be used as in the field, although there is the option of using different techniques which may be more accurate, such as conventional chemical analysis by titration or by the *colorimetric* method, in which reagents are added to the water sample, producing colour changes that are measured manually by eye using a colour chart or automatically, and more accurately, by spectrophotometer. For trace metals, a method known as *voltametric analysis* can be used. These laboratory methods, however, are not discussed further here.

By whatever means water quality sensors are deployed, the same types of sensor are used, and it is the sensors themselves that are the subject of the remainder of this chapter.

Temperature

Water temperature is important not only because it is a crucial climatic variable in the case of sea-surface temperatures but also because it affects the metabolism of organisms and the speeds of chemical processes. It also needs to be known in order to correct the readings of some of the other sensors.

It is possible to measure water temperature manually with mercury-in-glass thermometers, and this is indeed one of the methods still used for obtaining sea-surface temperatures from ships. But for most water quality applications, as well as most sea-surface temperatures, platinum resistance thermometers or thermistors are the usual choice. Their characteristics have already been covered in Chapter 3.

pH value

Acids and bases; a brief history   Acids and bases were known about long before their chemistry was understood – acids taste sour, bases feel soapy. In the eighteenth century, oxygen was discovered jointly by Joseph Priestley in England and Karl Scheele in Sweden. In 1777, Lavoisier in France interpreted the nature of oxygen as an element. He also suggested that acids were compounds that contained 'oxygen' (the name 'oxygen' means 'acid producer'). But this proved to be wrong when acids that did not contain oxygen were found. In 1816, Sir Humphry Davey proposed in England that all acids contained hydrogen, not oxygen. About 20 years later, the German chemist Justus von Liebig elaborated on Davey's idea by defining an acid as a compound which contained hydrogen that reacted with metals, producing hydrogen gas. But none of them attempted to explain the nature of acids at the atomic level. It was half a century later before Ostwald in Germany and Arrhenius in Sweden suggested that an acid was a compound that produced hydrogen ions in a water (aqueous) solution and a base or alkali was a compound that produced hydroxyl ions in water. That is, these compounds break up (dissociate) either completely or partly into charged particles called *ions* – positive *cations* and negative *anions*. Acids produce the hydrogen ion H^+ (a proton) whereas bases produce hydroxyl ions OH^- and a metal ion. Strong acids and bases are highly dissociated, weak ones are perhaps only dissociated to the extent of a few percent. For example, hydrochloric acid is completely dissociated, $HCl \rightarrow H^+ + Cl^-$. This concept was a great advance, and is still useful in most situations. Its weakness was that it applied only to aqueous solutions. Others, notably Bronsted in Denmark and Lowry in England, independently proposed in 1923 the more general definition that an acid is a species that can give up a proton and a base is a species that can accept a proton. At the same time the American chemist Gilbert Newton Lewis defined acids and bases in term of electron, rather than proton, transfer. However, for aqueous solutions, the concept that an acid is a compound that produces hydrogen ions and a base is a compound that produces hydroxyl ions in a water solution is the most useful.

Units and terms   The pH of an aqueous solution is a number expressing the relative amounts, or concentration, or activity, of hydrogen ions H^+ (protons) and hydroxide ions OH^-.

   Molarity is the most commonly used measurement of the *concentration* of a solution, being the number of moles of solute per litre of solution. A mole of a substance is the amount of that substance whose mass, in grams, is the same as its relative molecular mass. For example, a mole of molecular oxygen, O_2, has a mass of 32 g. A mole of any chemical species, such as an ion, always contains the same number of particles (6.02×10^{23}, Avogadro's constant). To make a solution of, say, 0.5 mol, half a mole of the solute is weighed and dissolved in the solvent (in the present case water), more solvent then being added to make the total volume one litre.

However, to be precise, it is the *activity* (or *effective concentration*) of the solute that is important rather than its actual concentration – that is, its tendency to take part in a given reaction. Because solutions do not always behave as 'ideal', some acting as if their concentration was more, or less, than it actually is, an *activity coefficient* is used to express the difference between activity and concentration. pH is actually a measure of hydrogen ion activity rather than the concentration of hydrogen ions. However, the coefficient approaches unity in dilute solutions and then the difference is more of an academic one from our point of view, but is mentioned because it can lead to confusion. The above terms were introduced in America in 1907 by Gilbert Lewis.

In 1909, Soren Sorensen defined pH as the negative logarithm (to base 10) of the molar concentration (H^+) of hydrogen ions. That is, $pH = -\log_{10}(H^+)$, or $(H^+) = 10^{-pH}$; the symbol pH stands for the power, or puissance, of the H^+ ions. For example, a pH of 3 indicates a molar concentration or activity of 10^{-3} hydrogen ions (10^{-3} mol).

(The logarithm of a number, to base 10, is the power to which 10 has to be raised to give the number. For example, $\log_{10} 100 = 2$, since $100 = 10^2$, and $\log_{10} 1000 = 3$, since $1000 = 10^3$, etc.)

The pH scale ranges from 0 (corresponding to a molar concentration of 1 mol H^+ – a strong acid) to 14 (corresponding to a molar concentration of 10^{-14} mol H^+ – a strong base). For example, a 0.01 mol solution of fully dissociated hydrochloric acid has a hydrogen ion concentration of 10^{-2} and thus a pH of 2. Water self-ionises slightly and this can be expressed as an acid–base equilibrium:

$$H_2O(l) + H_2O(l) \rightleftharpoons H_3O^+(aq) + OH^-(aq)$$

What this equation expresses is that since hydrogen ions consist of single protons, there is evidence to suggest that they attach themselves to water molecules, the resultant species being known as the hydroxonium ion, $H_3O^+(aq)$. pH is, therefore, more correctly, $-\log_{10}(H_3O^+)$. There being equilibrium between H_3O^+ and OH^- ions, the pH of water is 7, defined as neutrality. The dissociation is only slight and the water concentration is very high compared with that of the hydroxyl and hydroxonium ions.

Buffer solutions are solutions that buffer themselves against changes in pH by consuming added H^+ or OH^- ions through chemical association with other ions in the buffer. These solutions are used as references when calibrating sensors (see below) and are made up from capsules, typically for pHs of 4, 7 and 9, that are dissolved in water at the time they are needed. The capsules contain equimolar amounts of a weak acid or base and one of its salts. Although buffers are mostly associated with pH measurement, they are also made for many other ion species, such as calcium.

Ionic strength is a term that takes into account both the concentration or activity of ions in solution and the charge on the ions. The exact computation used is not important here, but it is useful to be aware of the term. (See the discussion later regarding the difficulty of measuring the pH of low-ionic-strength solutions). Conductivity measurements (see below) give a rough estimate of ionic strength.

The pH of unpolluted, natural river waters ranges from about 6.5 to 8.5, while rainwater is naturally slightly acidic (pH 5.6), because the carbon dioxide dissolved in it produces carbonic acid.

Measuring pH by indicators   The simplest way to get a rough indication of the pH of a solution is to use an indicator – a substance, usually a complex organic molecule (a weak acid or base) that changes colour at different pH levels. Litmus is the best known because it changes from red to blue as the pH crosses the neutral threshold of 7 from acid to base. However, many others exist, such as methyl green (with pH 0.2–1.8) and alizarin yellow (with pH 10.1–12.0). While these indicators are useful and cheap they cannot, alone, give a precise measurement, although they are normally used with titration methods that are precise. As titration is a laboratory technique it will not be included here. For field work, *ion-selective electrodes* are widely used to measure pH and, for automatic measurements in the field, they are the only practical option. Ion-selective sensors can also be used in titration, to detect the *equivalence point*, but this, too, is a laboratory procedure.

Measuring pH by ion-selective electrode   pH electrodes fall into the class of sensor known as specific-ion sensors or ion-selective electrodes (Durst 1969, Covington 1979); in this case it is the hydrogen ions that are to be measured. Figure 10.22 shows the electrode construction. Two electrodes are necessary, the ion-sensitive electrode and a reference electrode, each sometimes referred to as half-cells since both are necessary to form a complete cell. Although shown as two separate electrodes in Figure 10.22, in practice they may be brought together into one *combination electrode* (Fig. 10.23).

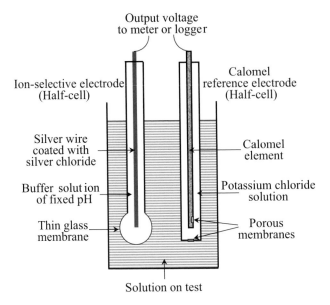

Figure 10.22. A pH electrode (on the left) develops a potential across its thin glass membrane; the magnitude depends on the concentration, or activity, of hydrogen ions in the solution on test. This voltage is measured relative to the fixed voltage generated by the reference electrode (on the right).

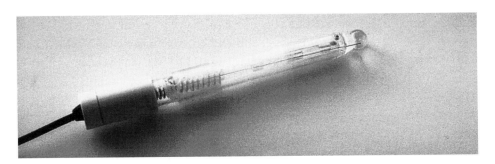

Figure 10.23. About 15 cm in length, this pH combination electrode integrates into one probe both the ion-sensitive electrode (the bulb) and the reference electrode (which makes contact with the solution via the small porous aperture just visible to the left of the bulb).

The ion-selective electrode is made of glass, the sensing part being a thin glass bulb at the end of the tube, this being placed in the water being measured. The bulb is filled with a standard acid buffer solution of known pH. Owing to the different concentrations of hydrogen ions in the fluid filling the bulb and in the water on test, a small voltage builds up across the thin glass membrane (in the hundreds of millivolts range). The composition of the glass determines to which ions the sensor responds. For example, glass electrodes are available that respond to sodium ions. Electrical contact with the buffer solution in the glass bulb is made via an electrode of silver coated with silver chloride, this acting as one side of the cell. The voltage across the glass membrane has to be measured relative to something and this is why a reference electrode has to be used. This electrode generates a fixed voltage and forms the second half of the cell.

The usual reference electrode for pH measurement is known as a *calomel* electrode, because it contains a central 'calomel element' made of a paste of mercury and calomel (mercuric chloride, Hg_2Cl_2) immersed in a saturated solution of potassium chloride (KCl), the paste being contained in a tube that has capillary holes in it to allow contact to be made with the solution (Ives & Lanz 1961). The outer tube containing the 'filling' solution, KCl, also has small capillary holes or a porous membrane allowing electrical contact with the solution on test. To avoid a build-up of ions inside the reference electrode from the sample under measurement, there has to be a very slow but definite flow of the reference electrode's filling solution into the sample – just sufficient to counter back-diffusion. Flow rates are in the order of 10–200 μl h^{-1}. The voltage at the point where the filling solution and the sample solution meet, the *liquid–liquid junction potential*, is stable, and it is this voltage against which the sensing electrode voltage is measured, giving an indication of the sample's hydrogen ion activity.

In 1889 the Nobel Prize winner, Herman Walther Nernst, introduced the Nernst equation, which relates the potential developed between the two electrodes of any electrical cell. The equation can be most conveniently expressed as

$$E = E_o \pm S \log a,$$

where E is the voltage measured between the sensing and reference electrodes, E_0 is a constant dependent largely on the type of reference electrode used, S is the slope of the mV versus pH curve, and a is the activity or concentration of the ion species being measured, in this case hydrogen ions. Because the slope of the curve is temperature dependent, it is necessary to compensate the reading for temperature, this being done manually or automatically.

The potential difference between the two electrodes may be measured manually with a portable meter displaying the measurement as millivolts or directly in pH units, or the voltage can be measured automatically and logged. By whatever means the voltage is measured, the high membrane resistance of the glass bulb, around 300×10^6 ohms, means that the amplifier circuit must have a very high input impedance so as not to load the cell, draw current and cause a drop in the detected voltage. Amplifiers, whether in a portable meter or a logging station, must have input impedances of at least 10^{12} ohms.

In the relatively pure waters of the natural environment, unlike those of industry and laboratories, the amounts of acid, bases and salts can be very low – as in normal rainwater for example. Measuring these is more difficult (Davison & Harbinson 1988) since the low ionic strength, and thus the low conductivity, of these solutions allows electromagnetic noise to be picked up rather easily. The use of lower-resistance glass membranes ($< 50 \times 10^6$ ohms) and low-ionic-strength buffers as electrode fillers, along with more complex (double-junction) reference electrodes, helps improve measurements. Natural waters also have a much lower buffering capacity, allowing for example the slow take-up of carbon dioxide from the atmosphere (if not already saturated with the gas) to appear as a drift in the pH reading.

Those planning to measure the pH of natural waters should, therefore, think twice before buying standard laboratory types of sensor and using them without precaution and awareness. Errors as high as 0.5 pH units can easily occur (Neal & Thomas 1985) if care is not taken. Some 'acid rain' work may well have ignored this factor. The accuracy of sensors used in laboratory conditions and in solutions of moderate ionic strength is in the order of ± 0.1 pH. The accuracy of measurements made in the field in low ionic strength water will depend very much on whether the above precautions are taken. Errors could be in the order of 1 pH unit.

Conductivity

The dissociation of compounds into charged ions in aqueous solutions (see the above discussion of pH) makes solutions conductive of electricity. If two electrodes in the solution have a potential difference applied between them, cations ($+$) migrate to the cathode ($-$), accepting electrons, while anions ($-$) move to the anode ($+$), giving up electrons. This transfer of electrons through the solution completes the electrical circuit. The conductive solution (*electrolyte*) obeys Ohm's law just as any other conductor does. As noted earlier, the conductivity of an electrolyte gives a fair indication of the ionic strength of the solution.

Units   The conductivity of a solution is defined as the reciprocal of the resistance in ohms of a one centimetre cube of the solution at 25°C across two opposite faces of the cube. The unit of conductivity is the reciprocal ohm per centimetre (mho cm^{-1}); mhos are also now known as siemens (S). As this unit is too large in practice for water conductivity, it is more usual to use the reciprocal megohm per centimetre, that is, the microsiemen per centimetre (μS cm^{-1}), or the millisiemen per centimetre (mS cm^{-1}), which, using the older names, are micromhos per centimetre or millimhos per centimetre (μmhos cm^{-1} or millimhos cm^{-1}). The microsiemen cm^{-1} is now the most common unit, the 'cm^{-1}' often being omitted leaving just μS.

However, a more basic definition of the conductivity of a substance is the reciprocal of the resistance of a cube of the material with sides of one metre, across two opposite faces, this being expressed as μS m^{-1}. As the resistance across the faces of a 1 cm cube of material is 100 times that across a 1 m cube, its conductivity is 0.01 of that of a 1 m cube. Thus, if a value of conductivity is, for example, 2 S m^{-1}, this would be equivalent to 0.02 S cm^{-1}. Converting to microsiemens (by multiplying by 10^6) gives 2×10^4 μS cm^{-1}.

Very pure water has a conductivity of around 0.05 μS cm^{-1}, distilled water around 3.0 μS cm^{-1}, with natural waters ranging from 35 to 500 μS cm^{-1}. In comparison, a 2% (by weight) solution of sulphuric acid has a conductivity of 93 000 μS cm^{-1}.

For saline water, the term *salinity* is used in place of conductivity, although it is still the conductivity that is being measured and by the same type of sensor. The units are also different, with parts per thousand (ppt) or parts per million (ppm), by weight, being used instead of units of conductivity. Seawater salinity averages about 34 grams per kilogram (or 34 ppt). This is equivalent to 34 000 ppm. As 1000 ppm of salt solution is equivalent to a conductivity of 2000 μS, the conductivity of seawater is in the order of 64 000 μS.

Conductivity sensors   Conductivity sensors consist of an electrode assembly designed to measure the resistance of a precise volume of water between two electrodes of precise area. Since, by definition, the conductivity is the reciprocal resistance of 1 cm^3 of the water, a sensor that has two electrodes of 1 cm^2 spaced 1 cm apart (or the equivalent), is said to have a *cell constant* of 1.0. Variations in the area and separation of the electrodes give different cell constants, for example 0.1, 1.0 and 10. Instruments are usually switched to cover different ranges, depending on application, the most common being from 0 to 10, 100, 1000, 10 000 or 100 000 μS. A cell constant of 1.0 will cover most of these ranges, with perhaps 0.1 being used for the lower end and 10 for the top. Accuracy ranges from $\pm 0.5\%$ to $\pm 2\%$ of full scale, although these values are for well-maintained, clean sensors.

Most conductivity sensors consist of an epoxy resin tube containing electrodes in the form of rings, flush-fitting on the inside of the tube. To minimise corrosion, the electrodes are made of stainless steel, tungsten or carbon or have a platinised finish. To ensure that a precise volume is measured, the electrodes are arranged as shown in Fig. 10.24. As when measuring soil moisture with gypsum blocks and humidity with resistive sensors, AC excitation has to be

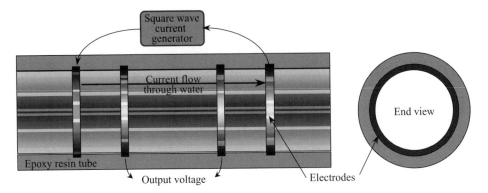

Figure 10.24. This section through a conventional conductivity sensor shows the four electrode rings, the outer two of which introduce a current through the water, the voltage drop across the resistance of the water being picked off by the two inner rings, which define a precise volume of water.

used to avoid polarising the sensor, a constant AC square-wave current being made to flow through the water between the two outer rings. This generates a voltage along the length of the water path, the voltage being related, by Ohm's law, to the resistance of the water and the current flowing. The two inner electrodes, precisely positioned to encompass the required volume of water, pick off the voltage, which is converted to resistance and thus to conductivity.

Because the electrodes can become dirty, an alternative design has been developed based on the magnetic coupling between two toroidal ferrite cores, parallel to each other and both wound with coils, as shown in Fig. 10.25. One coil is energised with an AC current, the resultant magnetic field inducing a voltage in the second coil. Water occupies the space through the centre of the two coils, the degree of coupling between them being proportional to the conductivity of the water; any increase in ion mobility causes a corresponding increase in the output from the detector coil. The response, however, is non-linear and temperature dependent, these being corrected by an on-board microprocessor. Accuracies of around $\pm 1\%$ of full scale are claimed, with the same ranges of conductivity as are available with the conventional electrode type of sensor.

Oxygen

Since animals living in water depend on the *dissolved oxygen* (DO), it is one of the most important water quality variables to be measured. The amount present depends on the balance between *inputs* of oxygen from the atmosphere and from photosynthesis by aquatic plants and *extraction* of oxygen by

Figure 10.25. By measuring the magnetic coupling between two adjacent coils wound on toroidal ferrite cores (in the circular head in the figure), the conductivity of the water can be measured without the need for any exposed electrodes.

respiration and as a consequence of the bacterial decay of organic matter (the *biochemical oxygen demand*, BOD), the latter sometimes resulting in a marked decrease in the amount of oxygen in water.

Units   Oxygen makes up 20.95% of the volume of dry air (that is, air containing no water vapour). It is only slightly soluble in water and the amount is dependent on the pressure of the gas and the temperature of the water. At one standard atmosphere (1013.2 hPa, or 760 mm of mercury) and a water temperature of 0°C, the maximum solubility of oxygen is 14.6 mg per litre (or 14.6 ppm). This falls to 13.5 ppm if the pressure is reduced to 700 mm (at 0°C) and to 7.7 ppm at one atmosphere if the temperature rises to 30°C. So under the normal range of pressure and temperature, there will be a maximum of between 7.5 and 15 ppm of oxygen dissolved in natural open water; there may be much less if the BOD is high. The concentration of DO is sometimes expressed as a percentage, 10 ppm of oxygen being the maximum concentration at standard pressure and temperature, and this is taken to be 100% concentration. Thus if the water contains 7.5 ppm (as at 31°C) the concentration is 75%, while at 0°C it is 146%. Meters are usually calibrated in ppm (mg litre^{-1}) or in per cent (0–20 ppm or 0–200%).

Oxygen sensors   The most commonly used oxygen sensor, or variations on it, is still the type designed by Mackereth (1964). In this design, oxygen diffuses through a permeable membrane of polypropylene of thickness 0.05 mm into an

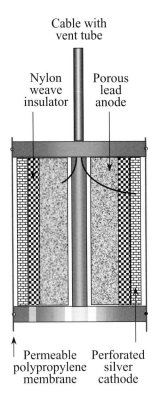

Figure 10.26. The cylindrical lead and silver electrodes of a Mackereth oxygen sensor are bathed in an electrolyte of potassium carbonate, the whole assembly being sheathed in an oxygen-permeable polypropylene tube 0.05 mm thick. One wire from the cable goes to the porous lead anode, the other to the perforated silver cathode.

electrochemical cell (Fig. 10.26) made up of a perforated cylindrical silver cathode, contained within the membrane and insulated by a nylon-weave insulator from a porous lead anode. The whole is bathed in a saturated solution of potassium carbonate (there are variations on this). The cell generates a current proportional to the partial pressure of the oxygen – about $10\,\mu A$ at 0.21 atmospheres and 25 °C.

The cell is calibrated by placing it in distilled water, stirred to ensure air saturation. When the current it produces is stable, this corresponds to the maximum concentration of O_2 that is possible at the prevailing barometric pressure and water temperature. By referring to a table relating temperature and barometric pressure to the maximum solubility of oxygen in water saturated with air, in ppm, a graph can be drawn relating sensor current to the maximum possible amount of DO in ppm. The graph is a straight line going almost through the origin. There is in fact a slight current of about 1% of the air-saturated value when no oxygen is present, and for precise work this current can be measured by immersing the sensor in de-aerated water (obtained by bubbling nitrogen through the water for five to ten minutes) and a more precise graph plotted.

Not only is the amount of oxygen that can be dissolved in water very temperature-dependent, but the current produced by the sensor also changes considerably with temperature, increasing by 2.5% per degree rise in temperature. To convert sensor current into ppm of oxygen, therefore, requires that the temperature also be measured, with the necessary adjustments being made either in the sensor electronics or in the computer software if the data are logged.

The current is usually converted to a voltage (by passing it through a resistor), so that it can be displayed on a millivoltmeter, such as a pH meter, for manual readings, or the readings can be logged. Accuracy ranges from $\pm 1\%$ to $\pm 5\%$ of full scale. But the accuracy also relies, as does the speed of response of the sensor, on the water being circulated over the sensor membrane effectively and continuously and on the membrane being clean. When these conditions are ensured, a sensor has a response time of about 30 seconds to reach 90% of a step change.

Biochemical oxygen demand BOD is of more concern to the operators of sewage treatment plants than to those working with the natural environment, and its measurement also involves a laboratory procedure. It is included here because it could, on occasion, be of use in natural waters. BOD is a measure of the use by bacteria of the oxygen dissolved in the water.

The measurement of BOD involves placing a sample of the water in an airtight bottle, the DO level having first been measured, typically with a Mackereth sensor. The sample is then incubated at $20\,^{\circ}\text{C} \pm 1\,^{\circ}\text{C}$ for several days. The DO content is then measured again and the consumption of oxygen expressed as ppm per litre over the period. Demand may vary over the period and this can be followed by measuring the DO level throughout, giving a measure of the rate of consumption.

Turbidity

Particles suspended in water range from small to large; the smaller are generally said to cause *turbidity*, while larger particles such as sand are described as *suspended sediments*. Generically, both smaller and larger suspended particles are often referred to as turbidity. Turbidity affects light transmission, fish eggs and life on the bottom of rivers; it is present in lakes and reservoirs and may have bacteria and pesticides attached to it. Land use and vegetation cover are the main determining factors of turbidity levels. How turbidity is measured depends on the amount of matter in suspension.

Measurement of turbidity   The turbidity of water is measured by the way it affects the passage of light through it, either by its attenuating effect on a direct beam or by its sideways or backward scattering of the beam. Methods using scattering are often referred to as using the 'nephelometric' principle, nephology being the study of clouds.

For low levels of turbidity, the amount of light scattered at 90° to a light beam is measured. For higher levels, the attenuation of the direct beam is sensed – usually based on a one metre or quarter metre path length, sometimes folded back on itself, using a prism, to shorten the instrument. For yet higher turbidities and for suspended sediments (such as mud and sand), the back-scatter from angles ranging from 180° to 140° is measured. Figure 10.27 illustrates these options, while Fig. 10.28 shows an actual sensor.

In the past, filament lamps had to be used as the light source, and some designs still use them, although they take a large amount of power. In place of filament lights, many present-day instruments use light emitting diodes, which are more stable, longer lived and less power demanding; some use xenon-gas-filled tubes. These latter two sources can also be pulsed, and this reduces power consumption further as well as providing an AC signal in place of a DC one, which improves electronic stability. The pulse rates used are in the 5 and 10 Hz range. The optical windows of these instruments can become dirty, but this, and changes in the light-source intensity as well as in detector sensitivity, can be compensated by measuring absorption over two paths of different lengths in the water, cancelling out the variations. This method can also be used with filament lights, which slowly age. Some instruments include automatic cleaning, either ultrasonic or by a mechanical pump-like piston which periodically moves past the optical windows. Some designs combine 90° scatter and direct transmission in one instrument, thereby allowing a distinction to be made between light losses due to dissolved matter in the water (giving the water its colour) and particles in suspension; both attenuate the light beam but only the particles scatter it.

The wavelength of the light used ranges from the visible to both the infrared and ultraviolet. (UV from a xenon light source is used for the measurement of fluorescence in detecting, for example, chlorophyll, although this is not considered here.) IR has the advantage that longer wavelengths do not penetrate far below the water surface (being absorbed), thereby reducing the interference from natural light. IR transmission is sufficient, however, over the short distances used in instruments, to measure sideways-scatter or back-scatter

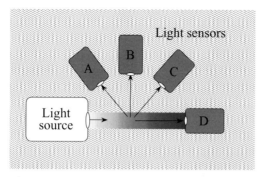

Figure 10.27. By detecting the degree of attenuation or the amount of scattering of a beam of light, the concentration of suspended particles can be measured. A: for very high turbidity, the back-scatter is measured. B, C: for low levels, 90 degree or forward-scatter is sensed. D: for intermediate levels, attenuation of the direct beam is used.

over a few centimetres. Filtering the light, so as to sense just the wavelengths used, further reduces the interference effects of natural light. In deeper water, there is little penetration of light from the surface and so no effect on the signal from natural light.

Units and calibration   Calibration is normally done using formazin – the turbid precipitate formed by mixing a solution of hydrazine sulphate with a solution of hexamethalinetetramine ($CH_2(6)N_4$), the combination producing a precipitate with a mix of particle sizes similar to those found in naturally turbid water. It is prepared as a stock solution equal to 4000 FTU (*formazine turbidity units*, also known as NTUs – nephelometric turbidity units). In use, the stock solution is diluted to the required levels, from very low values, 1–10 FTU, up through the hundreds to one or two thousand. (Once diluted, the solution has a useful life of only 24 hours; the stock solution has a life of one year.) Because hydrazine sulphate is a carcinogen, safer, long-life alternative calibration solutions using copolymer particles have been developed, but they are more expensive and do not have such a wide range of particle sizes.

   The absorption and scattering of light in water depends very much on the size distribution, shape and colour of the suspended particles, which vary from white chalk to dark mud, and it would be impossible to calibrate an instrument to encompass all of these. Indeed, one of the problems with formazin is that it is a white precipitate and there can be a 5:1 ratio, or more, between the reflective and scattering properties of formazin and those of natural dark particles. To be certain of good turbidity measurements, therefore, it is necessary to calibrate sensors in a range of suspensions that

Figure 10.28. In this particular turbidity sensor, two LEDs (lower window) emit IR light into the water, the IR detector (upper window) measuring the light scattered forward from the suspended particles.

match the water being measured. Over-reliance on factory calibrations or formazin standards could lead to large errors.

Low turbidities, measured by $90°$ scatter, cover the range 0 to 100 FTU (NTU). Higher values, measured by direct-path absorption, fall in the 100 to 500 FTU range. Although there is considerable overlap, anything much greater than 500 FTU is usually considered as suspended sediment rather than turbidity and is expressed in ppm or mg l^{-1}, values ranging from 5 to 5000 ppm for mud and 100 to 100 000 ppm for sand.

References

Ackers, P., White, W. R., Perkins, J. A. & Harrison, A. J. M. (1978) *Weirs and Flumes for Flow Measurement*. John Wiley & Sons, Chicester.

Barnes, H. H. (Jr) & Davidian, J. (1978) in *Hydrometry, Principles and Practices*, ed. Herschy, R. W., pp. 149–203. John Wiley & Sons.

Benson, M. A. (1968) Measurement of peak discharge by indirect methods. WMO No. 225.TP 119, Technical Note No. 90.

Biswas, A. K. (1970) *History of Hydrology*. North-Holland, Amsterdam and London, ISBN 7204 8018 3.

Bowden, K. F. (1956) The flow of water through the Straits of Dover related to wind and differences in sea level. *Phil. Trans. Roy. Soc.*, **248**, 517.

Corbet, D. M. *et al.* (1945) Stream-gaging procedures. US Geological Survey (USGS), Water Supply Paper 888.

Covington, A. K., ed. (1979) *Ion-selective Electrode Methodology, Vols. I and II*. CRC Press.

Cox, R. A. (1956) Measuring the tidal flow in the River Humber. The Dock and Harbour Authority, July 1956.

Dalrymple, T. & Benson, M. A. (1967) Measurement of peak discharge by the slope-area method. Chapter A2 in Techniques of Water Resources Investigations, Book 3. USS.

Davis, J. M. (1979) Calibration of multi-path ultrasonic stations. In *Proc. Ultrasonic River Gauging Seminar*, pp. 27–38. Water Research Centre and Department of Environment Water Data Unit.

Davison, W. & Harbinson, T. R. (1988) Performance testing of pH electrodes suitable for low ionic strength solutions. *Analyst*, **113**, 709–13.

De Pitot, H. (1732) Description d'une machine pour mesurer la vitesse des eaux courantes et le sillage des vaisseaux. Mémoires de l'Académie Royale des Sciences, 363–76.

Durst, R. A., ed (1969) Ion selective electrodes. Special Publication 314, National Bureau of Standards.

Faraday, M. (1832) *Phil. Trans. Roy. Soc.*, 1832, 175.

Frazier, A. H. (1969) Dr. Santorio's water current meter, circa *1610*. *J. Hydraulics Division, ASCE*, **95**, 249–54.

Gilman, K. (1977) Dilution gauging on the recession limb: (1) constant rate injection method. *Hydrol. Sci. Bull.*, **22** (3).

Herschy, R. W. (1979a) Site selection and specification of configuration. In *Proc. Ultrasonic River Gauging Seminar*, pp. 6–15. Water Research Centre and Department of the Environment Water Data Unit.

Herschy, R. W. (1979b) Uncertainties in ultrasonic flow measurement. In *Proc. Ultrasonic River Gauging Seminar*, pp. 103–5. Water Research Centre and Department of the Environment Water Data Unit.

Herschy, R. W. (1995) *Streamflow Measurement*. Chapman & Hall, London.

Herschy, R. W. & Fairbridge, R. W. (1998) *Encyclopedia of Hydrology and Water Resources*, p. 800. Kluiver Press.

Herschy, R. W. & Newman, J. D. (1974) Electromagnetic river gauging. In *Proc. Symp. River Gauging by Ultrasonic and Electromagnetic Methods*, UK Water Research Centre and Department of the Environment Water Data Unit.

Horton, R. E. (1907) Weir experiments, coefficients and formulas. US Geological Survey, Water Supply Paper 200.

Ives, D. J. G. & Lanz, G.J., ed. (1961) *Reference Electrodes, Theory and Practice.* Academic Press.

Leliavsky, S. (1951) Historic development of the theory of the flow of water in canals and rivers. *Engineer*, **1**, 466–8.

Leliavsky, S. (1965) *River and Canal Hydraulics.* Chapman & Hall, London.

Limerinos, J. T. (1970) Determination of the Manning coefficient from measured bed roughness in natural channels. US Geological Survey, Water Supply Paper 1898-B.

Mackereth, F. J. H. (1964) A sensor for dissolved oxygen. *J. Sci. Inst.*, **B**, 38.

Neal, C. & Thomas, A. G. (1985) Field and laboratory measurements of pH in low conductivity natural waters. *J. Hydrol.*, **79**, 319–22.

Riggs, H. C. (1976) A simplified slope-area method for estimating flood discharges in natural channels. *US Geological Survey J. Res.*, **4**, 285–91.

Walker, S. T. (1979) Where is ultrasonic gauging appropriate? In *Proc. Ultrasonic River Gauging Seminar, pp.* 3–5. Water Research Centre and Department of the Environment Water Data Unit.

Water Research Association (1970) River flow measurement by dilution gauging. WRA Bulletin T. P. 74.

11

Data logging

Before the development of modern data loggers in the 1960s, the only means of automatically recording measurements of the environment was on paper charts, either mechanically or on electrical strip-chart recorders with electrical sensors. It was the arrival of solid-state electronics, in particular its ability to operate digitally, that enabled computers and data loggers to be developed. Both have greatly enhanced the way in which the natural environment can be measured, indeed they have revolutionised it.

The construction of a data logger

The schematic of Fig. 11.1 shows each main section of a data logger. With the development of large-scale integration on one *integrated circuit* (IC) chip, and of the microprocessor, many of these functions are now carried out on a single IC, supported by a range of peripheral chips such as serial data communicators, memory access controllers, counters and clocks (Fig. 11.2), although even many of these are now on one single chip. However, to explain the functioning of a logger, it is useful to keep the boxes separate. Indeed they were, in reality, physically separate (until the development of the larger ICs in the 1980s), the first loggers using individual transistors, resistors and capacitors with wires interconnecting them. Today, a small number of ICs, mounted on printed circuit boards, perform all of the functions required – in a reduced space, at reduced cost, with increased reliability and with very low power requirements.

Interfacing

Most sensors are analogue, producing electrical signals that vary continuously in response to the variable being measured. Few analogue signals are

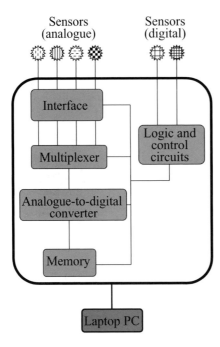

Figure 11.1. A logger's interfacing circuits first standardise the wide variety of analogue signals that sensors produce, the processed signals then being switched sequentially by a multiplexer to an analogue-to-digital converter. Pulse signals, such as those from raingauges and anemometers, bypass these processes, being already digital. As a final step, all the readings are stored in memory. A separate memory is programmed to hold user instructions, controlling how the logger processes the sensor signals and stores them, this process, and the subsequent downloading of the logged data, usually being via a laptop PC.

immediately useable in their raw state and it is the role of the logger *interface* to process them into a standard form, usually to one volt (sometimes five volts) full scale. There are a number of kinds of conversion that need to be done to meet this requirement, but the majority are covered by a few simple processes.

Potentiometer sensors   Potentiometer sensors can be made to generate the required voltage signal by applying a precision reference voltage across them; the 'wiper' picks off a proportion of this as the signal, depending on its angular position, which is related to the amplitude of the variable. The reference voltage need only be applied for the few milliseconds required for the logger to take a reading, and so the power consumed is minimal. An error in the reference voltage will cause an extra error in the reading, additional to those generated by the sensor itself.

Figure 11.2. A present-day data logger will typically have all the integrated circuit chips, and the few necessary discrete components such as resistors, on one printed circuit board with tracks on both sides, through-hole plated. Connections to the sensor inputs, to the PC (for programming and download-ing of data) and to the power supply are made via the plugs and sockets seen along the lower and upper edges of the board. The two chips with the round windows are EPROM memories, the chip components being visible through the windows for erasure by UV light. This is a complete logger, able to handle eight analogue and two pulse circuits. It consumes only a few microamps in its quiescent state.

Low-voltage sensors   Where sensors generate a voltage directly, but not of sufficient amplitude, as in the case of radiation sensors, for example, which have outputs in the millivolt region, an amplifier must be used to increase the signal level to the required one volt full scale. Like so much electronics following the development of the IC, amplifiers today are contained on one chip, indeed one chip may contain several. There are those that handle AC signals, such as for audio and video applications, and those that amplify DC signals, such as those generated by most environmental sensors. Before ICs were available, it was difficult to amplify DC voltages accurately, the signals first having to be 'chopped', to convert them to AC for amplification; they were then converted back to DC signals afterwards. While this is still the most precise method, for most environmental applications there are now many low-cost, low-power, IC operational and instrumentation amplifiers which perform extremely well, and it is generally these that are used in data loggers.

Any error in the zero or gain (amplification) settings of the amplifier will be added to those of the sensor, and this can be affected by the voltage rails which power the amplifier and by the means through which the gain and zero are adjusted. Adjustment may be by fixed or trimmable resistors external to the IC chip, although loggers now tend to use programmable-gain amplifiers, in which the gain is set by logger software, ensuring good accuracy. Amplifier temperature coefficients are typically ± 10 ppm K^{-1}.

DC amplifiers are either single- or double-ended (*differential*). Most loggers today use the latter, which can be operated in either mode to suit the particular application, the differential mode being used when the sensor output is electrically 'floating', neither side being connected to earth (or common or ground).

Most of the water quality specific-ion sensors, typified by pH, have very high internal resistances and the amplifiers used with them must have a much higher input resistance than normal (at least 10^{12} ohms), so as not to load the sensor and reduce the signal.

Resistive sensors   If a sensor, such as a platinum resistance thermometer (PRT), does not produce a voltage signal directly, it must be derived. In the case of a PRT, the usual method is first to generate a millivolt signal by making the sensor one arm of a bridge circuit; this is followed by amplification. While manually operated bridges are balanced by hand to give a null output on a meter, in an automatic system it is usual to allow the bridge to go out of balance and to log the unbalanced output voltage, after amplification.

When two or more stages of signal conversion are necessary, as here, there are more potential sources of error, including errors in bridge-excitation voltage as well as in the amplifier, these being in addition to the sensor errors and sensor exposure errors (discussed in Chapter 3 in the sections on electrical thermometers and exposure of thermometers). The overall total error will depend on how the individual errors combine, some adding and some subtracting depending on the circumstances, which may change with time.

Some resistive sensors cannot have a DC excitation voltage applied across them because they become polarised. In such cases, for example soil-moisture gypsum blocks and some resistive humidity sensors, an AC excitation voltage must be used. Otherwise the principle is the same, although the final output from the interface must be converted to a DC voltage in the required 0–1 volt range. In such cases, the logger usually supplies the required AC excitation internally.

Capacitive sensors   A few sensors, such as those for humidity and pressure, vary in capacitance as the variable changes. The value of the capacitance is usually very small (picofarads) and so it is not possible to connect them to the logger over a long cable, which will have its own (much larger) capacitance. Such sensors have the interfacing electronics within a few centimetres of them, usually supplied by the manufacturer of the sensor in the actual sensor housing, although the logger may have to supply the power to operate them.

Pulse-producing and digital sensors   Pulse signals, such as those from rain-gauges and wind-speed sensors, are dealt with differently, there being no need for analogue-to-digital conversion. Instead, they go directly to the logic circuits of the logger, as described in the subsection below on logger programming.

A few sensors, such as shaft encoders, produce serial-binary or parallel-binary signals (such as the Gray code). Loggers must include a special interface card to handle these, which enable most of the normal logger processing operations to be bypassed.

Multiplexing

Some loggers accept just one input, such as for a stand-alone raingauge, but most are multichannel with provision for 2, 4, 8, 16 or more analogue inputs, and one or more pulse-counting sensors, as well as serial or parallel digital signals. When there are two or more analogue inputs, the logger samples them sequentially, switching (*multiplexing*) rapidly from one to the next. While this was once done electromechanically by reed switches, today the switching is done electronically: each of the outputs from the interface circuits is switched, one after the other, to the next stage of the logging process, which is the *analogue-to-digital* converter (ADC). In fact the multiplexer and ADC are now usually on the same chip, probably combined with the processor and also with programmable amplifiers.

Analogue-to-digital conversion (ADC)

Digital terms   It is useful to clarify the meaning of three words used frequently in discussing digital data. A *bit* means a binary digit (1 or 0). A *byte* is a group of bits treated as a unit, typically involving 8, 10, 12, 16 or 32 bits, and is usually, but not always, shorter than a *word*, which is a string of bits, usually upwards of 16. But 'word' is also used more loosely in place of 'byte' (Pitt 1982).

It is also useful to look at the various *number codes* in common use. The numerical examples below are taken from Horowitz & Hill (1984), from which more detail can be obtained.

Decimal numbers   A decimal number (base 10) is the familiar form of numbering used in everyday life. It needs ten symbols (0–9) to express a number. An example follows; the subscript indicates that base 10 is being used:

$$137_{10} = 1 \times 10^2 + 3 \times 10^1 + 7 \times 10^0$$
$$= 100 \quad + 30 \quad + 7$$

Straight binary numbers   A binary number (base 2) needs just two symbols to express a number (0 and 1). This is useful when working with computers and loggers because 0 and 1 are easily represented by two positions of a switch (on or off) or by two ends of a voltage range (for example, $+5$ and -5 or $+5$ and 0). To see the relation between decimal and binary numbers, consider the following:

$$1101_2 = 1 \times 2^3 + 1 \times 2^2 + 0 \times 2^1 + 1 \times 2^0$$
$$= 8 \quad + 4 \quad + 0 \quad + 1$$
$$= 13_{10}$$

Octal and hexadecimal codes   Binary numbers tend to be long, and so it is convenient to break them up into shorter lengths. Octal numbers (base 8) are binary numbers broken up into groups of three binary digits, each group requiring eight symbols (0 to 7) to express a number. Consider the number

$$835_{10} = 1101000011_2$$

Broken up into groups of three, starting from the least significant bit, this number becomes

$$1 \ 101 \ 000 \ 011$$
$$= 1 \quad 5 \quad 0 \quad 3$$
$$= 1503_8$$

Hexadecimal numbers (base 16) are obtained by breaking a binary number into groups of four digits, each group having values from 0 to 15 and requiring 16 symbols (0–9 together with A–F) to express a number. Thus one counts 0123456789ABCDEF. As an example,

$$707_{10} = 1011000011_2$$

Broken up into groups of four binary digits, this number becomes

$$10 \ 1100 \ 0011$$
$$= 2 \quad C \quad 3$$
$$= 2C3_{16}$$

Hexadecimal is useful because of the convention of organising numbers into bytes (8-bit groups), usually as 16- or 32-bit words; a word becomes 2 or 4 bytes in hexadecimal format. The problem with this code is getting used to strange numbers and doing arithmetic with them.

12- and 36-bit words were common in earlier computers and this suited the octal format better, with its divisions into groups of three binary digits. But with 16- and 32-bit words hexadecimal is more useful.

Binary-coded decimal (BCD)   This number code converts each decimal digit into binary form – hence the term BCD – using a 4-bit group for each decimal digit. Using a previous example,

$$137_{10} = 0001\ 0011\ 0111\ (BCD)$$

This is not the same as the binary version of this number, which would be 10001001_2.

BCD is wasteful of bits since each group of four binary digits could express a number 0 to 15, while BCD never represents a number greater than 9. However, BCD is useful if a number is to be displayed in decimal, since all that is necessary is to convert each BCD group of four binary digits to the equivalent decimal number. Many electronic ICs are available to do this – BCD decoders, drivers and displays. So the BCD format is the one used most commonly for the input and output of numerical information.

The Gray code   In Chapter 5, in the subsection on automatic wind direction sensors, and in Chapter 10, in the subsection on sensing water level by float, the concept of a shaft encoder was introduced, in which the sensor turns a shaft that rotates a disc encoded digitally. It would be a simple matter to produce such a disc with a pattern on it, coded in straight binary, that could be read optically, producing a parallel output indicating the state of each of the bits of the number representing the angular position of the shaft. But, rather than use a straight binary code on the disc, it is preferable to use what is known as the Gray code. This code has the property that, unlike in binary, only one of the bits changes at any one time in going from one state to the next. The reason for preferring this is that in straight binary format several bits can change when going from one state to the next. In practice it is very difficult to ensure that the mechanical alignment is good enough to guarantee that all the bits that should change, do change together and at exactly the same instant. If they do not, large errors in the output can occur. The Gray code avoids this by changing the single least significant bit that gets it to the new state. Table 11.1 shows the decimal numbers 0–9 in the Gray code.

Table 11.1. *The Gray code*

Decimal	0	1	2	3	4	5	6	7	8	9
Gray	0000	0001	0011	0010	0110	0111	0101	0100	1100	1101

Gray codes can have any number of bits; just four are shown here. It is possible to convert from Gray to binary and back using suitable electronic gate circuits (Horowitz & Hill 1984). In the case of shaft encoders used with float-operated water level sensors, their outputs can be chosen to be Gray coded or can be converted to BCD (for ease of display) or straight binary. But apart from special cases such as BCD and Gray codes, computing and logging circuits work in straight binary.

Resolution   In Chapter 1, the term 'resolution' was defined as the smallest change in a variable that could be measured by a sensor (Fig. 1.2). The same considerations also apply to the logging of the measurements: the larger the number of bits used in ADC, the finer the resolution obtained. Table 11.2 shows the resolutions obtained for different word lengths. The larger the number of bits, the finer the resolution obtained. Thus, if an 8-bit ADC is used, the resolution obtained is 0.4% of full scale; this might be adequate for, say, humidity, which cannot be sensed more accurately than this. However, 12-bit word lengths will give a resolution of 0.02%, which can be of value in measuring, say, river level, where this can be done by sensors and where it can therefore be useful to have such sharp resolution.

A justification sometimes made by users regarding high-resolution loggers is that they want to look at small changes, not at absolute values, and that the extra resolution provided by 12 bits helps them do this. Although they may have a case, care is still needed so as to be certain that the small changes revealed are real and not due to random noise in the system or to sensitivity or zero drifts within the sensor or power supply. Also, when the data get into the public domain a warning about the fact that high resolution might not mean high accuracy is rarely printed alongside the data. All data should bear an indication of their origin, of the type of instrument used to collect them and a guide as to how accurate they are.

The electronics of analogue-to-digital conversion   How electronic circuits operate is not the concern of this book; however, it is useful to look at the principle of analogue-to-digital conversion. There are several means of achieving it, as follows.

Table 11.2. *Bit numbers and resolution*

Bit number	Bit value	Total value	Resolution (%)
1	1	1	100.00
2	2	3	33.33
3	4	7	14.29
4	8	15	6.66
5	16	31	3.23
6	32	63	1.59
7	64	127	0.79
8	128	255	0.40
9	256	511	0.20
10	512	1023	0.10
11	1024	2047	0.05
12	2048	4095	0.02

In *successive approximation*, the conversion is performed by a trial-and-error process. Using a *voltage comparator* circuit, the voltage equivalent to the most significant bit (MSB) is compared with that of the signal. If the signal is the greater, that bit is left on (1), if less it is turned off (0). This process is repeated for each next-most-significant bit in turn until the least significant bit (LSB) is reached, each step progressively homing-in on the correct reading. It is a quick and accurate method and probably the most widely used (Fig. 11.3).

In a similar technique, the voltage to the comparator increases, in steps equivalent to the LSB, until it is the same as the input voltage, when the process stops; the number of steps is counted and represented as the binary value of the voltage. Because it has to step through each possible voltage from zero upwards, this can be a (relatively) slow process.

Lesser-used circuits are the *dual slope* (or *charge balancing*) method, and the fastest, and most expensive, known as the *parallel conversion* technique.

Memory

Brief history   Once converted to digital form, the sensor measurements are stored in memory. In the early to mid 1960s, quarter inch magnetic tape was used as the storage medium, both reel-to-reel and cassettes. After a few years, a change was made to the new *compact cassette*, Fig. 11.4 (Strangeways & Templeman 1974). A few loggers used half inch computer tape in large cassettes. All these tapes were played back on special readers into the new mainframe computers of the day, such as Digital's PDP8. Tape had many good characteristics – compact cassettes were available worldwide in every town, they were

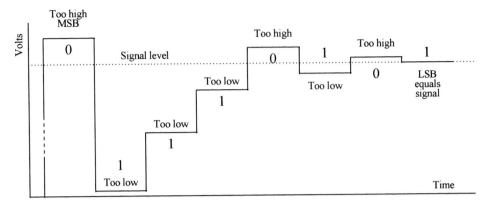

Figure 11.3. In the successive approximation method of analogue-to-digital conversion each bit, starting with the most significant bit (MSB), is compared with the signal and if greater is set as 0, if less as 1. The 8-bit word illustrated, therefore, represents the binary number 01110101 (or 117 in decimals).

Figure 11.4. Prior to the availability of low-cost, high-capacity, low-power-consumption RAM, magnetic tape, usually in the form of compact cassettes as in the figure (right front), was widely used and served well from the late 1960s up until the late 1980s, although now replaced entirely by solid state memory. This particular logging system has a separate interfacing unit (left), each card containing a circuit for the conversion of a sensor signal into a 5 volt standard for multiplexing and analogue-to-digital conversion; these processes are carried out in the right-hand case. The rechargeable battery is seen in the lid.

cheap (£1), could store between 50 000 and 100 000 8-bit bytes, could be posted, the data on them were secure, the tapes could be reused and no power was required to maintain the memory. Their main disadvantage was that they needed mechanical tape-drives and these could be a problem in the field, particularly at low temperatures. Downloading of data was also time consuming, requiring the tapes to be played back at a rather low speed. Nor was tape-replay possible in the field. (It is useful to be able to look at the logged data while still at a field site as a check on the correct functioning of the station.) During the 1980s, solid state memory became available, ending three decades of tape dominance.

Types of solid state memory   Memory is used in a logger both to store the programmes for operating it and to store the measurements made by the sensors, the former requiring memories that do not need frequent erasure and reprogramming, while the latter must be cleared and reused many times over.

EPROM, or electrically programmable read only memory, can be programmed and then erased for reprogramming. Erasure is by UV light through a quartz window in the chip, the UV acting directly on the components of the IC chip (Fig. 11.2). It was designed for occasional erasure, and today is widely used to hold the programmes that operate the logger. At one point in the 1980s, however, before RAM (below) became cheap, some loggers did use EPROMs to store the data but the UV erasure was a disadvantage and very frequent reuse over a period could eventually lead to some data corruption.

EEPROM (E^2PROM) or electrically erasable and programmable read only memory, is similar to EPROM but erasure is done electrically, which is more convenient than by UV light. They too are meant for occasional reprogramming.

RAM, or random access memory, is intended for the temporary storage of data and for frequent reuse, making it ideally suited to the storage of data in a logger. RAM is usually of complementary metal-oxide semiconductor (CMOS) fabrication. MOS transistor construction allows very high packing densities, while CMOS logic circuits use complementary MOS transistors, that is a pair of opposite type (n–p–n and p–n–p), which has the advantage of low power consumption (Pitt 1982). Many loggers now use removable (RAM) memory cards with capacities up to 4 Mbyte.

Logger programming

Early loggers were not programmable. The number of channels and their functions were selected by the addition or removal of interface cards from a rack (Fig. 11.4), sensor ranges were adjusted with a screwdriver by turning a trimpot, the analogue clock was set by hand and the interval at which readings were taken was adjusted by switches. But with the ability to incorporate a microprocessor, logger function became much more flexible and under the control of software.

Today, the user can specify the following: how many channels are to be used; the type of sensors that will be connected; the ranges to be covered; the *scan interval* (5, 15, 30, 60 seconds – the frequency at which readings are taken and stored temporarily); the *logging interval* (the frequency, for example hourly, at which values are stored in memory for later collection, which can be different for each channel); whether any processing is required, such as the averaging of the readings taken at the scan intervals or the logging of extremes measured during the logging interval, spot readings at the logging interval or units to work in (metres, %, °C); how the data are to be output to the PC and plotted on-screen; the date and time at which logging is to start. Details of the field site (name, or latitude and longitude) can also be included. Whether data should continue to be stored in memory when the memory is full (new data overwriting the old), or whether logging should stop, can also be specified. Details of communication between the logger and the PC via their serial ports also need to be set up, specifying such things as baud rate (the rate at which data are transmitted, in bits per second), parity bits (for error checking) and byte length. Although most loggers are programmable in the above way via a PC (temporarily attached to the logger's serial port), some have a built-in liquid-crystal display and simple keyboard that allow interaction with the logger.

There are two options for handling pulses, which is especially useful in the case of a tipping bucket raingauge. The total number of pulses during the logging interval can be logged, giving the rain total for the period, or the time and date of each pulse can be logged – in the *event mode* – the logger coming out of its quiescent state momentarily to log each event as it occurs. This gives an indication of rain intensity, as well as allowing totals to be calculated. Which mode is used depends on requirements. For example, in an arid environment it is wasteful of memory to log six months of hourly zeros, so the event mode may be preferable. However, if there is frequent high rainfall the event mode may take up too much memory, since there will be a great many events to log. It is also possible to use a *delta function*, in which the logger logs only when a variable has changed by a preset amount. Logging might also be initiated only when an event such as the tip of a raingauge bucket occurs.

Often all these instructions to the logger can be set up on a PC in advance of use, stored on the PC's disc and finally loaded into the logger with just a few key-strokes, cutting the amount of programming work needed in the field, which often has to be done under difficult conditions and under the pressure of time.

Power supplies

CMOS logic circuits consume little power, and with loggers in a quiescent state most of the time (when only microamps are drawn) overall power consumption is very low indeed. For these reasons, field stations can be operated unattended for long periods at remote sites where there is no mains power. There are four common ways of powering a logger, as follows.

Replaceable cells, normally of the alkaline type, can operate a station for up to a year and can be bought worldwide. *Rechargeable nickel–cadmium batteries* are an alternative, but *lithium batteries* are increasingly being used as they can power a logger for up to five years. *Solar panels* (Fig. 11.5), charging *lead–acid* (sealed-gel) batteries, are also now commonplace and can power a station indefinitely, even in higher latitudes. Except under special circumstances the logger battery also powers all the sensors that need power and all the interface circuits. Just occasionally, however, special external interface circuits may be necessary, for example with water quality sensors, and these may need a dedicated power supply of their own.

Collecting the logged data

After a day, a week, a month or a year, the logging station will be visited and the recorded data retrieved. Although removable memories are a feature of some designs, retrieval is most usually done by downloading the data to a PC's floppy or hard disc, via the serial port of the PC and logger (Fig. 11.6). This is a fairly rapid process taking but a few minutes, and the data can then be returned to base. An advantage of on-site downloading to a PC is that it is then possible, while still at the field site, to check the performance of the logger – perhaps by plotting the data on-screen. But this requires some training of the operator and mistakes can be made. Straight memory swapping is simpler and requires less skill.

It is usually possible to make the choice whether to reprogram the logger at each visit after downloading the data or to let it run on. If the former is selected, all the old data are cleared from memory; in the latter case the old data remain, the new data following on after them. An advantage of reprogramming at each visit is that the logger clock is automatically resynchronised with the PC's clock, while if the logger is left to run on then any drifts in its clock setting will

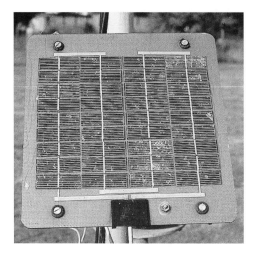

Figure 11.5. To avoid the need to change batteries and to allow long-term operation, many environmental instruments are now powered by lead–acid-gel batteries charged by a solar panel, the one illustrated being large enough to operate most forms of remote instrument, even in higher latitudes.

Figure 11.6. The most usual way of collecting data from a logger at a field site is to connect a laptop PC and to download the data to disc. Here data are being collected from a tipping bucket raingauge; the logger, normally housed in the box, is standing next to, and connected to, the PC on top of the box. The raingauge cover has been removed for testing, making the tipping bucket visible.

Figure 11.7. Here, two layers of 5 cm thick expanded-polystyrene insulation, contained in a wooden box, protect the logging and satellite telemetry units contained within it. This was used at the BAS Faraday base in Antarctica in connection with experiments carried out by the Institute of Hydrology, on experimental cold-regions sensors (Fig. 13.9). While anything other than extremely thick insulation cannot prevent the equipment from experiencing the mean temperature, insulation does protect against the extremes (including very high temperatures in hot climates).

accumulate with time. An advantage of letting the logger run on is that the old data are still there if the downloading process has failed.

Protection from the environment

Loggers are typically housed in hermetically sealed boxes of plastic or metal, with sealed plugs and sockets for the connection of the sensors and the PC. A bag of silica-gel crystals is often placed inside the box to keep it dry. It is also usual to protect the logger further, by housing it in a hut at the top of a stilling well for river level measurement or in a small extra enclosure at an AWS.

Loggers can thus be fully protected from dampness and most physical damage – but they are not so easily protected from temperature extremes. The inside of an unventilated box or hut exposed to the sun can become hotter than the surrounding air, even if the box is white, and in situations where air temperatures can reach $+45\,°C$ or more some additional shielding from the sun may be advisable to prevent too high a rise in temperature. At the other extreme, in polar regions, temperatures can fall to $-45\,°C$, or lower. The temperature specifications of ICs, batteries and other components vary considerably, typical ranges for ICs being 0 to $+70°C$, -20 to $+85°C$ and -45 to $+85°C$. Nickel–cadmium and lithium batteries can work over temperatures from -40 to $+60°C$, although the former may not retain all their capacity at the extremes and can suffer from a high self-discharge at raised temperatures. Logger manufacturers quote temperature specifications ranging from $-20°C$ to $+50$ or $+60°C$, with just a few covering -40 to $+70°C$. By use of military-specification chips (at increased cost) the range can be extended, but often normal chips also will operate well beyond their specified limits, although tests are necessary to confirm this. By burying the logger in the ground or in a snowpack, or by using well-insulated boxes (Fig. 11.7) temperature extremes can be reduced, but usually the logger must simply be able to work over the full range.

References

Horowitz, P. & Hill, W. (1984) *The Art of Electronics.* Cambridge University Press, Cambridge UK. ISBN 0 521 29837 7.

Pitt, V., ed. (1982) *The New Penguin Dictionary of Electronics.* Penguin Books, ISBN 0 14051 074 5.

Strangeways, I. C. & Templeman, R. F. (1974) Logging river level on magnetic tape. *Water and Water Eng.,* **178**, 57–60.

12

Telemetry

Reasons for telemetering data

Telemetry is the transmission of data from one point to another. If data are needed in real time they must be telemetered, for example for weather forecasting and flood warning. Telemetry also has two significant advantages over *in situ* data logging, even if the measurements are not required in real time: the cost of visiting field sites to collect data is saved and the failure of field stations can be detected – months of data could be lost if a logging station failed soon after a visit. Logging is best suited to applications where stations are within relatively easy access or where the loss of some data is not a serious problem.

The general process of telemetering data is sometimes referred to generically as *system control and data acquisition* (SCADA), although the term applies more strictly to management applications – where not only are data acquired from a remote location but remote control is also exercised back. A dam managed from a distant control-room, for example, is a more appropriate use of the term SCADA than is the one-way collection of environmental data.

The structure of a telemetry system

Figure 12.1 is a schematic of a telemetry system, showing its main subdivisions into sensors, logger, modem, communications link and a PC at the base station. This basic arrangement is similar for all telemetry systems although it will differ in detail, mostly depending on the communications link used. A telemetry system is in effect a logging station with a communications link appended and with a remote base station to receive the transmitted data. The front end of such a field station is composed of sensors, identical to those used at a logging station, and a unit that performs the same functions as a logger, even if it is not called such – interfacing, multiplexing, analogue-to-digital

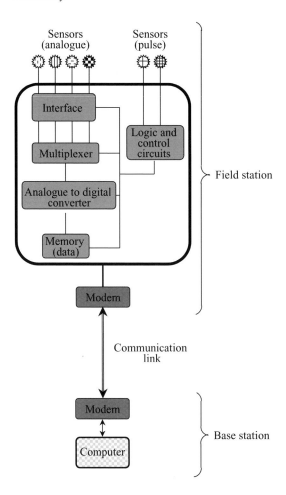

Figure 12.1. A telemetry system is a logging station with a communications link added.

conversion and memory. Previous chapters have been concerned with the sensors and with logging; all that needs to be addressed in this chapter is the communications link.

Modems

Loggers and computers express the digital '1' and '0' as DC voltage pulses; typically '1' corresponds to $+5$ volts and '0' to -5 (or 0) volts. DC pulses cannot, however, be transmitted over communications links, and must be converted into *voice frequency tones*, which *can* modulate a communications link. This is done using a *modem* (MOdulate–DEModulate) – for modems also perform the reverse function of converting tones back to DC pulses when they are received at the far end of the link. The process by which a modem converts digital 1s and 0s into tones is known as *frequency-switch keying* (FSK), there being one frequency for 1s and another for 0s.

The base station

The base station is generally at a central office where the data are required, either for immediate and automatic input to a computer model for a real-time application or for archiving for later use. The hardware of the base station comprises the communications link and a PC, or perhaps a larger computer. The PC performs several functions: calling up (*polling*) the field stations, receiving their returned data and processing them.

The communications link

Certain basic principles apply to the way in which communication is carried out, whatever the communications link. If the link is able to send data both ways at the same time, such as over a telephone line, it is termed *full duplex*. With two-way communication the base station can *poll* the field stations at any time and instruct them to send their data. If the link can send data both ways, but not at the same time, it is known as *half duplex* and this is more than adequate for telemetry. Some links, however, can only carry information in one direction, and these are known *simplex* channels. Polling is not then possible, and the base station has to wait for the field station to send its data at set times.

There is much benefit to be had from being able to poll field stations from base, since they can then be interrogated as often or as little as necessary and, in the case of a network of stations, each station can be called at different times and at different intervals. With simplex channels there is no option but for the field station to send its data at preset intervals and for the base station to wait for the transmissions. This, however, may be perfectly adequate.

Telephone, ground-based radio or satellites can be used as telemetry links. Which is most suitable will depend on circumstances.

Terrestrial communication links

Telephone lines

The *public switched telephone network* (PSTN), as used for normal telephone voice communication, also now carries digital signals from faxes and PCs. Baud rates, more usually now referred to as bits per second, as high as 56 000 are now commonplace and rising. Before the introduction of electronic exchanges and fibre-optic links, the telephone network was not as reliable as it now is. Telephone lines are now very well suited to telemetry and, because communication is duplex, field stations can be polled; nevertheless, there remains some possible truth in the view that they are unreliable, since the last stretch of a telephone link to a field site may be vulnerable to damage by wind or flooding, and so particular care is needed in this final step of the link. The cost of installing lines to the remoter sites may also be high, or impractical, in which case an alternative will need to be found. But where telephone lines can access a site and where the country's network is modern and reliable, telephone links are very attractive and are widely used for telemetry. Where a mobile telephone service is available these can also be used to telemeter data, but unfortunately the signal level is often too low in many remote areas, which is where it would be most useful.

The only hardware that is required to convert a logging station into a telephone-telemetering station is a modem, the cost of which is small. The cost of calls will depend on the usual factors: the number of transmissions made, their duration and distance and the time of day. Where real-time data are not required, measurements can be stored and sent at night or at the weekend, but where this is not possible, low-cost scheduling may not be an option. Indeed, the ongoing cost of calls can be a deterrent to the use of telephone links, in which case radio may be preferred.

VHF and UHF radio

Where telephone lines are not practicable, technically or because of cost, but where distances are not great, terrestrial line-of-sight *very high frequency* (VHF) and *ultra-high frequency* (UHF) radio is the usual alternative, although satellite links are increasingly being used even for these local situations. The VHF band extends from 30 MHz (10 metres wavelength) to 300 MHz (1 metre), merging at the upper frequency into the UHF frequencies, 300 MHz to 3 GHz (10 cm). UHF, in particular, behaves like light: the waves do not significantly bend beyond the horizon, restricting their range to about 30 km

Figure 12.2. Although there can be refraction beyond the horizon, UHF and VHF radio offer essentially only line-of-sight links, even if nothing more than the curvature of the earth intervenes. Hills and buildings often obscure the view and make links shorter, although, conveniently placed, they can be used to extend the range, by elevating a base or field station antenna.

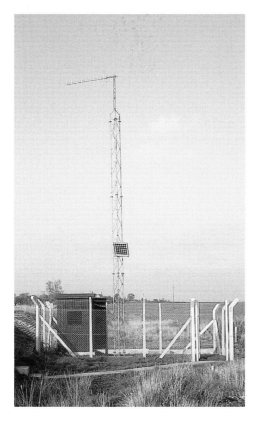

Figure 12.3. The line-of-sight distance covered by this river-gauging station's UHF radio link has been increased by installing the antenna on a mast.

(Fig. 12.2). There may be some slight refraction beyond the line of sight, especially with the lower VHF frequencies and in certain weather conditions, but such communication is not reliable. High antenna masts (Fig. 12.3) can extend the range, as can high ground, but high ground is only useful if it is advantageously positioned; hills more often restrict the range by obstructing the path of the beam. Repeaters can be used (Fig. 12.4), but they add to the complexity of a network, increase cost and reduce reliability.

Figure 12.4. If a UHF link cannot cover the required distance in one span, a repeater can be used, usually sited on a hilltop as is this water tower; many organisations make use of this facility.

Because these radio bands can be focused into narrow beams using a Yagi antenna, the power required is small and 5 watts is generally sufficient; power levels are also restricted by law to about this level. Interference from other transmitters is reduced by the directional nature of the beam. These bands are very crowded and it can be difficult to get a licence to use them. Despite line-of-sight limitations and a crowded spectrum, VHF and UHF provide good communication and are the only ground-based alternative to telephone lines and in consequence are widely used. They also have the advantage that the calls are free. In the long term, the cost of operating the transmission link (the cost of telephone calls, or of using a satellite) can be the greatest cost in operating a telemetry network. The UHF frequencies are the more commonly used because their higher frequency gives a greater bandwidth and thus a greater number of channels is available.

Recently a UHF method using low-power transmitters (half a watt), have become available; these can cover distances up to 15 km. Their main advantage is that they do not require a licence for operation (in the UK at least).

As with telephone lines, two-way communication (half-duplex in this case) is possible with UHF and VHF links, but because the radio receiver must be in operation, waiting for a polling call from base, too much power may be consumed if the site is reliant on solar power or replaceable batteries. A similar problem also occurs with those satellite links that allow duplex communication. For this reason, networks operating in the *simplex mode* have some attraction, since they are simpler and do not need to have a receiver at the field stations. Such systems transmit the data from the field stations either at set intervals or when an event occurs, such as the tip of a raingauge bucket.

However, even though the small amounts of data usually involved can be transmitted in under a second, there remains the possibility that two stations will transmit at once, thereby resulting in the loss of both messages. Provided that the number of field stations is not too large, nor the transmissions too frequent, the chance of coincident transmissions is usually acceptably small. But if for any reason data are not received for a period of time, owing perhaps to a fault at the base station or at a repeater, in such a simplex system these data will be lost in real time. It may be possible to retrieve them later by visiting the field stations affected, provided that the stations include a data logger for *in situ* recording, but if the data are needed in real time, they will not be available. Only polling can minimise such losses, since it allows the stations to be called when the fault is corrected, and the lost data accessed.

In a polling network, as a compromise to save power it is possible to call the field stations just at set times, say during a 15 minute window every six hours, to reprogramme the stations if necessary or to call up data missed in the intervening six hours; between times the stations simply transmit at set times or when an event occurs. To achieve this, the field station is programmed to switch on its receiver at pre-arranged times to await any calls from base.

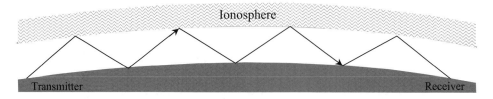

Figure 12.5. The ionosphere reflects HF radio waves (3 to 30 MHz), permitting multiple reflections between it and the ground and thus enabling communication over great distances. The C-layer of the ionosphere extends from 40 to 60 km above the earth's surface, the D-layer from 60 to 90 km, reflecting the lower part of the band, the E-layer from 90 to 150 km, reflecting the medium frequencies, and the F-layer from 150 to 1000 km, reflecting the high-frequency end of the band. However, fading and interference make it a less than ideal communications link and it is little used for telemetry today.

HF radio

Before satellite telemetry was available, *high-frequency* (HF) radio was the only channel for telemetering data over a long distance where there was no telephone network. It is included here more for completeness than in any likelihood of its being used, now that satellites are available.

More commonly known as *short-wave radio*, HF radio frequencies extend from 3 MHz (100 metres wavelength) to 30 MHz (10 metres). Radio waves in this band are propagated by successive reflections between the ionosphere and the ground (Fig. 12.5), thereby spanning great distances with a power of only 50 watts. But the ionosphere moves up and down during the day, as the seasons change, and as a result of solar activity, so the signals suffer from fading and from interference from other transmitters. The waves can also travel by several slightly different paths (multipath propagation) and this distorts the received waveform because they arrive at slightly different times. While these problems can be reduced by using multiple-antenna and multiple-frequency techniques, these are very expensive. However, before satellites, HF was successfully used for telemetering flood warning data in Brazil (Strangeways & Lisoni 1973; Fig. 10.4). Today, however, satellites do the job more reliably, and about the only thing to be said for HF now is that, like UHF, calls are free; perhaps also it has the advantage that negotiations do not have to be made with any agency.

Meteor-burst communication

Micrometeorites, the size of particles of sand, enter earth's atmosphere continuously, leaving small ionised trails that last a few seconds. These trails will

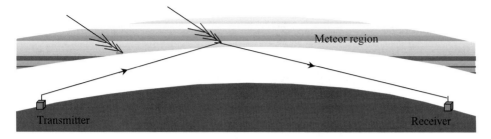

Figure 12.6. The short-lived ionisation trails produced by micrometeorites can be used to reflect radio waves in the 30 to 200 MHz band over distances up to 2000 km but, owing to cost and technical complexity, such meteor-burst communication is only rarely used for environmental telemetry projects.

reflect radio transmissions in the 30 to 200 MHz range and this property has been used as a means of long-distance communication. The most effective band is between 30 and 50 MHz (the low end of the VHF band).

Field stations using this technique direct their (rather large) Yagi antenna at that part of the sky where an ionised trail, when it occurs, will reflect the signal to the distant base station (Fig. 12.6). The base station keeps calling the field station until an ionised trail occurs at the right place and the right time, reflecting the call to the field station. Averaged over a year, the *waiting time* is typically about two minutes. When the field station receives the call it replies, and if the ionisation is still active the base receives the message; if not, the base keeps calling until it gets a reply without errors. The main problem is the very high cost of the base station (£100 000+), which is complex (Fig. 12.7). But it also has practical problems. Signal levels reflected off meteor trails are extremely low and a compromise is necessary between transmitter power (100 watts being the minimum) and the gain (or sensitivity) of the antenna. If the antenna gain is high, that is it has more dipole elements, it directs its beam more narrowly and so with more concentrated power, but it then reaches fewer ionised trails. If, at the other end of the link, the receiver gain is increased beyond a certain threshold to detect weaker signals, it may respond to man-made or natural electromagnetic noise.

The maximum distances that can be covered are up to about 2000 km from base to field station, although this is adequate for all but the largest networks. While meteor-burst communication has occasionally been used for environmental data telemetry, this use is restricted because of cost and complexity, although, recently, cheaper methods have been developed (Schafer & Verner 1996) and hiring a channel is sometimes an option.

Figure 12.7. The large size of the antenna necessary for meteor-burst com-munication (because of the low VHF frequency used) is apparent from this substantial tower at a base station. Four antennae are necessary to achieve 360 degree coverage.

Satellite communications

Since the 1970s, satellites have been launched for a variety of environmental applications, the foremost of which are meteorological, notably for the collec-tion of images, particularly of clouds but also for the measurement of variables such as temperature, rain, humidity and evaporation. While *in situ* surface measurements are the main topic of this book, remotely sensed observations are closely related, and the next chapter illustrates what can be measured remotely and how the two sets of data complement each another. However, in addition, the same satellites have also been equipped with the capability of relaying data telemetered from remote field stations to a distant base. It is this capability that is the topic of the remainder of the present chapter.

Because environmental–weather satellites have the dual capability of image-generation and telemetry, it is convenient to look at the characteristics that affect both functions first, and one of the important distinguishing features of satellites is their orbits. There are two types of orbit into which satellites with telemetering, remote-sensing and communications capabilities are normally placed – geostationary and polar.

Orbits

Geostationary orbits   Satellites with an orbital distance of 35 900 km from the earth's surface circle the earth exactly once a day. If placed in such an orbit around the equator and rotating in the same direction as the earth, the satellite appears stationary in the sky, and is known as *geostationary-equatorial* (Fig. 12.8).

Polar orbits   Satellites that orbit the earth over the poles can have any altitude required because they do not have to keep step with the earth's rotation. Meteorological satellites in polar orbits fly at an altitude of about 850 km, giving an orbital period of around 100 minutes. The plane of the orbits of polar satellites used for imaging and telemetry is kept facing the sun, the earth rotating beneath it, and these satellites are thus known as *sun-synchronous*; a satellite passes over a different swath of ground at each orbit, gradually scanning the whole of the globe, with the sun always at the same angle at each pass (Fig. 12.8). Polar orbits that are not truly polar are also used, in which the angle of the orbit is not vertically north–south. Some communication satellites, such as the new Orbcomm network, use a mixture of orbits, equatorial, polar and skewed, to achieve full global communication at all times.

Orbits compared   Geostationary orbits present the satellite with the same view of the earth (*footprint*) all the time, the whole circle of the globe being visible (Fig. 12.9). Immediately beneath the satellite, the earth's surface is seen square-on, but in all other directions the view is increasingly oblique, to a point where there is no view at all. In particular, the polar regions can never be seen from geostationary satellites.

Polar orbits, in contrast, give an ever-changing view of the surface, the satellite sweeping out a path up and down the earth and over the poles; each part of the earth is viewed more or less vertically downwards and thus to the same resolution over the whole globe and across the whole image. Pixels

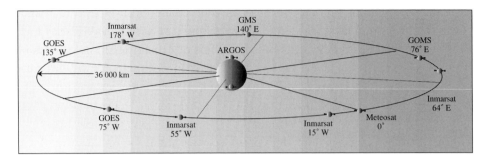

Figure 12.8. The principle satellites used for telemetering environmental data include Meteosat, the two US GOES satellites (that at 75° west being GOES East and that at 135° west being GOES West), the GMS satellite of Japan, the Russian GOMS and the three ARGOS satellites in polar orbit. In addition the four Inmarsat satellites are used for telemetering environmental data, although at present to a much lesser extent.

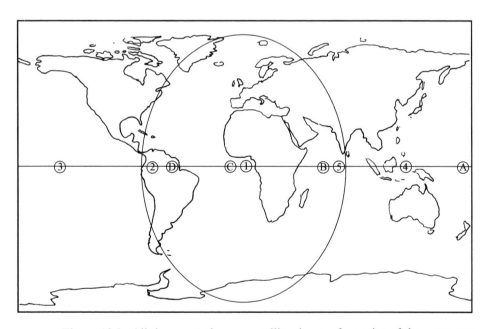

Figure 12.9. All the geostationary satellites have a footprint of the same area and shape. In this figure, for clarity, only that of Meteosat is shown, centred on point 1. The others are centred on the points indicated: 2, GOES (East); 3, GOES (West); 4, GMS; 5, GOMS. The Inmarsat satellites are centred at A, B, C and D.

representing areas as small as 30 m square or even smaller are achievable. However, repeat images may be many hours apart, even days or weeks.

From a telemetering field station's viewpoint on the ground, geostationary satellites appear at a fixed point in the sky, their angle above the horizon becoming progressively lower the higher the latitude of the observer and the further away the observer is longitudinally from vertically beneath the satellite. Higher than about 75° latitude, the satellite's angle above the horizon is too low for reliable telemetry, although this depends partly on the nature of the terrain. In valleys, for example, the cut-off latitude can be lower, perhaps no higher than Scotland. Because the satellite's position is fixed, telemetering field stations can use directional antennae to concentrate the radio wave into a narrow beam towards the satellite, thus making the best use of limited power. By this means, the 35 900 km to the satellite can be bridged with as little as 5 to 10 watts.

In contrast, a polar-orbiting satellite passes over a telemetering field station quickly and disappears from view until the next pass. So while a field station can telemeter via a geostationary satellite at any time, a polar satellite is only available for about ten minutes or less during each orbit. Further, while polar satellites pass over the poles at every orbit, at any point on the equator a polar satellite is only in range for telemetry about four times a day – even though it crosses the equator the same number of times as it crosses the poles. With an orbital time of 100 minutes there are around 14 passes a day at the poles. Thus telemetry via a polar satellite is possible only during the approximately 10 minute overpass, at a repeat interval dependent on latitude. Because the satellite passes across the sky, field-station antennae cannot be directed in any one direction and instead have to be omnidirectional, directing the radio emission evenly in all directions. This reduces the signal level received at the satellite but, because polar satellites are around 25 times nearer than those in geostationary orbit, the signal level is adequate, although a higher radiated power of 20 watts has to be used.

As well as being a disadvantage, satellite movement relative to the ground can also be used to advantage since it results in a Doppler shift in the radio frequency received at the satellite from the telemetering field station (see the discussion, later in this chapter, of the Argos system).

The family of satellites   At the present time, most of the satellites used for telemetering environmental data are the weather and environmental satellites, put into orbit primarily as collectors of remotely sensed images. Those in geostationary orbit are Meteosat (Europe), GOES East and GOES West

(USA), GMS (Japan) and GOMS (Russia). The Inmarsat network of four commercially operated geostationary satellites encircles the globe, offering an alternative to the weather satellites. Polar-orbiting satellites operated by the US National Oceanic and Atmospheric Administration (NOAA) are used for telemetering data through the French Argos system. All these various satellites are shown in Fig. 12.8. The Russian Meteor weather satellites are another set in polar orbit and Eumetsat (the organisation managing the European Meteorological Satellite) plans to have a polar orbiting network of three satellites in operation by 2003, which will produce high-definition images (Eumetsat 1997).

In all cases there are spare satellites in orbit as standbys, for use if an operational satellite should fail, and there is an on-going programme of launches to replace satellites nearing the end of their life and also to improve performance through new developments. Failures can occur and the need to fine-tune the orbits of the satellites from time to time requires on-board fuel, which gradually gets used up over the years; this gives satellites a finite lifetime.

There are also many national satellites, such as India's geostationary Insat, America's Domsat and Brazil's SCD-1 satellite (in equatorial orbit with an inclination of 25°, giving coverage of all Brazil and of all the world's tropical and subtropical countries). The last was designed specifically for environmental data collection. None of these can be dealt with in this short treatment. The situation will change considerably over the next decade, with new systems being launched, such as the Orbcomm network of 36 satellites in low orbit offering duplex communication and e-mail capability; this has just gone into service. Indeed the present weather satellites should be viewed as just the beginnings of satellite telemetry. Meteosat is a convenient satellite to explain the general principles of satellite environmental telemetry, for it incorporates many of the features found in the other weather satellites.

Meteosat

Meteosat is operated by the European Space Agency (ESA) and managed by Eumetsat, its mission control centre being situated in Germany in Darmstadt and its primary ground station at Fucino in Italy (Eumetsat 1996).

Data collection platforms   Field stations telemetering via satellite are usually referred to as data collection platforms (DCPs), but the term is used somewhat loosely and may refer to the whole field station or to just the electronic package containing the transmitter. Whichever satellite is used, the field stations are similar to a terrestrial radio telemetering station – they are made up of sensors, logger and communications link, the only difference being that in this case the transmitter is directed at a satellite. A DCP performs identical functions to any other telemetering field station, making the measurements, multiplexing through them, converting them to digital form and storing them ready for transmission, as Fig. 12.1 shows (Strangeways 1990).

Transmission time-slots Telemetry via the geostationary weather satellites is (generally) one-way only (simplex), the DCP transmitting its measurements at fixed times, these being allocated by the satellite operator. An *address*, a string of (usually eight) numbers to identify the DCP, is also allocated to each station. Time-slots are usually spaced at three-hourly intervals, to keep in line with the main use of all the weather satellites – weather forecasting by the NWSs –, data being collected every three hours. More frequent time-slots are technically feasible (since the satellite is in constant view), but this needs special approval. Each time-slot has a duration of one minute, during which the station's measurements are transmitted to the satellite at a slow baud rate, typically 100 bits per second, along with housekeeping data such as battery voltage, DCP address and perhaps indications of equipment performance. This slow baud rate restricts the amount of data that can be telemetered, but it is usually adequate for most environmental applications. Two consecutive time-slots can be arranged when data amounts require it.

Although transmissions are at set time intervals, the field station can nevertheless make measurements at whatever interval the user cares to choose, such as every 15 minutes, these being stored ready for transmission at the prescribed time. For many uses this is adequate, but if events are liable to unfold rapidly, such as in the case of flash-floods, three-hourly intervals may not be sufficient. It can also happen that occasional transmissions fail to get through, for one reason or another, and these data are lost in real time (although they may be retained in on-site memory for later retrieval). This can mean that no data are received for six hours, which is perhaps unacceptable for some real-time applications.

As DCPs transmit at set times, their clocks have to be precise, to avoid transmissions overlapping adjacent time-slots. To guard further against such interference between two stations, there is a buffer of one minute between each time-slot, allowing a certain amount of clock-drift. Transmissions out of time-slots are viewed with considerable disapproval by the satellite operators and immediate remedies requested. As DCPs are often sited at remote locations, which are usually difficult to reach, high reliability and good clock stability are important so that visits to readjust clocks or to reprogramme a system that has malfunctioned do not have to be made too often.

There is an exception to out-of-time transmissions in the form of *alert* or *random* messages, which can be sent at any time – on a special channel free of fixed-time-reporting DCPs. Because two DCPs might send an alert message at the same time, they send them several times, each DCP having a different repeat time.

Polling DCPs   It is possible to poll DCPs, but not directly by the user from the base station. Instead, calls are initiated by the satellite operator, from the *ground station*. The call message contains the DCP's address and all DCPs (that have receivers) receive the call, only the one with that particular address responding. Such DCPs must be equipped, however, with a receiver, and these have to be powered-up all the time to await a call, which requires more power and makes the station much more complex and expensive. The great majority of DCPs are self-timed, transmitting at three-hourly intervals.

Meteosat currently has 66 normal channels for DCP telemetry, but the situation is always in flux (the Eumetsat web page is http://www.eumetsat.de). Channels are divided into *international* and *regional*, the former for DCPs that move from area to area, such as those on ships or planes, and the latter for fixed stations.

DCP hardware   A field station may take the form of a data logger connected to a DCP, the DCP containing just the transmitter, some logic and a small memory (Fig. 12.10); the logger and DCP communicate through their serial ports. In the simplest of cases the logger may be completely omitted, a sensor being connected directly to the DCP; an example of this is a telemetering rain station (Fig. 12.11) (Strangeways 1985, 1994, 1998). DCPs of this simpler type need to be programmed by the operator only to the extent of setting the clock: the time is input by the operator via a laptop PC or by a simpler dedicated *synchroniser unit* (Fig. 12.12). The time at which transmissions are to be made and the DCP's address are programmed into the DCP at manufacture, in these simpler DCPs, and these can only be changed by changing an EPROM memory chip.

The DCP and logger combination consume little power and most of the time are in a quiescent mode, coming into action only briefly to log the sensor readings, perhaps every 15 minutes, and to transmit the accumulated data, generally every three hours. Because the average power consumed is small, stations can operate using solar power – even in higher latitudes in winter.

Other DCPs are more sophisticated, containing all the elements, logger included, in one integrated package. Assembled in modular form with individual plug-in cards or rack-mounted units, the design becomes multipurpose, it

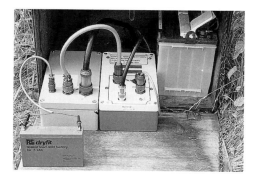

Figure 12.10. A data collection platform (DCP) of the simpler type (central) is here connected to a data logger (left) that records measurements from an AWS of the type shown in Figure 6.5. The car battery (right) powers the DCP, the small battery (left) the logger.

Figure 12.11. Although the simpler DCPs of the type illustrated in the previous figure do not have a logging capability, they can accept direct inputs from sensors producing suitable analogue voltages or pulses, in this example from a tipping bucket raingauge in the pit (centre). The Yagi antenna is directed at Meteosat from this Institute of Hydrology site in central Wales.

then being possible to select a combination of cards to suit needs. Thus it is possible to construct a station able to be operated as a telephone or a UHF ground-based telemetry system or as one able to telemeter via satellite; or it can perform simply as a logger. A DCP of this type is programmed by the user in much the same way as a logger, using a PC (Fig. 12.13). The precise time is input automatically from the PC's clock, but the times of the transmission slots, the DCP's address, the format the data should take (such as the SYNOP

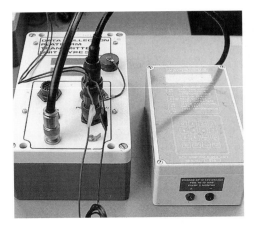

Figure 12.12. A DCP of the type illustrated in Fig. 12.10 (and in this figure to the left), requires little setting up. What programming is required is done using a 'synchroniser' (right), the main function of which is to set the DCP's internal clock precisely, to the second, to avoid transmitting out of the allotted time-slot. In addition the synchroniser can perform diagnostic checks on the DCP and on the sensors and initiate test transmissions into a dummy load. A laptop PC could perform the same functions as the synchroniser.

or CREX codes – see below) and the units in which the data are to be expressed must all be prescribed by the operator via the PC. This makes programming a more complex matter than with the simpler DCPs. Some such DCPs have built-in keyboards and displays that can be used, rather than a separate laptop PC (Fig. 12.14), and some incorporate a *global positioning system* (GPS) that automatically resets the clock regularly.

From DCP to ground station   The DCP radio transmitter operates in one of the bands between the frequencies 401 and 402 MHz; these are spaced at 3.0 kHz intervals, with a power of between 5 and 40 watts, the actual power depending on the type of antennae used. In the case of fixed stations the signal can be beamed at the satellite using a Yagi antenna (Fig. 12.11), in which case 5 watts is sufficient, but with a moving station the antenna must be omnidirectional and the radiated power must be at the higher level to achieve an adequate signal strength at the satellite. Upon arrival at the satellite, Meteosat retransmits the signal, at a frequency of around 1675 MHz, to the *ground segment* (Fig. 12.15). The ground segment is focused on the mission control centre at Eumetsat's HQ in Darmstadt in Germany, the primary ground station being in Fucino near to Rome; the two are connected by high-speed links via a commercial satellite, with a ground-based link as a back-up. At Darmstadt the data are quality controlled and then disseminated by two

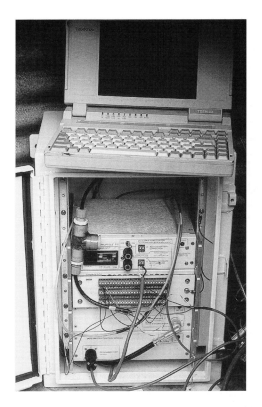

Figure 12.13. Although the basic DCPs illustrated in the previous figures can be connected to a data logger to allow a greater variety and number of sensors to be operated, DCPs of the type illustrated here perform the dual role of DCP and logger. They are also capable of more complex programming, such as being set up to telemeter the data in the SYNOP or CREX codes and to input their address number and details of the time-slots. Programming is achieved using a laptop PC (seen on top of the DCP case) in the same way that a logger may be programmed (Fig. 11.6).

routes – over the *global telecommunications system* (GTS) and back to the satellite for retransmission to the user (Eumetsat 1995a, b).

The global telecommunication system   The GTS is a communications channel interconnecting all the world's NWSs for the interchange and dissemination of meteorological data for weather forecasting. Data put onto the network must be presented in an internationally agreed WMO code, such as SYNOP. Whereas SYNOP is intended primarily for reporting meteorological measurements from land stations, there are other codes, such as SHIP for observations from sea, HYDRA for hydrological stations and the newer CREX code, which has provision for encoding both meteorological and hydrological measure-

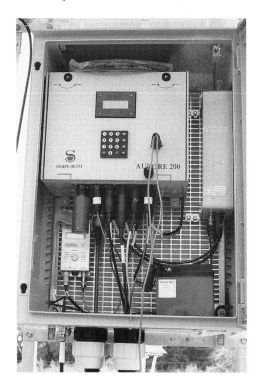

Figure 12.14. In a further variation on DCP design, illustrated here, the laptop PC is replaced by a built-in display and simple keyboard, although a PC can also be connected for more complex requirements, allowing less specialised staff to attend to the DCPs. This particular DCP also incorporates a GPS receiver that allows it to synchronise its clock regularly, and to confirm its position. It also has a removable back-up memory module (seen beneath the case at the left).

ments (WMO 1988). Programmable DCPs (Fig. 12.13) allow these codes to be set up by the user via a PC to suit the variables being transmitted, but this is quite a specialised matter. The simpler DCPs and loggers may not be capable of this and may have to be set up at manufacture.

Data sent over the GTS can only be received if access to the GTS is available, and this is generally only possible in co-operation with an NWS. This can be something of a disadvantage to all except operational meteorologists and is a drawback to the GTS when it is used for anything other than meteorology.

The DCP retransmission system, and Wefax images   In the case of Meteosat, however, there is an alternative to the GTS, for as well as being circulated over the GTS the data can be retransmitted from Meteosat for direct reception by

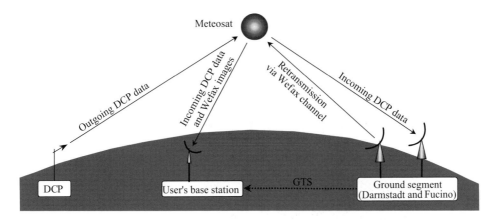

Figure 12.15. Typically every three hours, the field-station's DCP transmits its data to Meteosat during a one-minute time-slot, the data being received in real time at the ground segment. Here the data are disseminated either over the global telecommunication system (GTS) or are retransmitted over the Wefax channel for direct reception by the user's base station.

the DCP operator. To understand how this is done, it is necessary to digress into how Meteosat's remotely sensed images are collected and disseminated, for they are closely linked with data transmission. This digression, however, also provides some necessary background for Chapter 14, on remote sensing.

Meteosat's images are generated by an on-board multispectral radiometer (see Chapter 14), the images being transmitted digitally to the ground station in Fucino, Italy, and then relayed to the European Space Operations Centre in Darmstadt, where they are processed.

For immediate use, the images then follow two paths back via the satellite for retransmission to users. The first channel carries high-quality digital data for reception by what are known as *primary data user stations* (PDUSs). These need receiver dishes typically two to three metres in diameter (more about PDUSs in the next chapter). It is the second path, which carries the images in the form of an analogue signal known as Wefax (Weather facsimile) that is of concern here, for this channel also carries the retransmitted DCP data (Fig. 12.15). Image definition is lower than in the case of PDUSs, but the signal is simpler to handle (Eumetsat 1998). These transmissions are received on *secondary data user stations* (SDUSs), which not only receive the Wefax analogue images but also, interspersed with them, the retransmitted DCP data. Transmissions to SDUSs can be received on small dishes, typically 1.5 m diameter (Fig. 12.16), because the signal level is high, allowing low-cost reception of the images (Harris 1996). The signal is down-converted from its received frequency of 1694.5 MHz to 137 MHz at the dish, allowing cables of up to 100 metres

Figure 12.16. A small dish antenna is sufficient to receive the retransmission of DCP data via the Wefax channel from Meteosat. Unfortunately this very useful facility is not available from any of the other geostationary weather satellites, although direct reception is possible from the polar-orbiting Argos satellites under certain circumstances.

length between the dish and the receiver. The format taken by the Wefax transmissions is relevant to the reception of DCP data and so needs explanation, as follows.

Wefax images build up line by line, as in a normal TV picture but much more slowly, taking $3\frac{1}{2}$ minutes rather than one twenty-fifth of a second, the incoming data being handled by a PC and displayed on its monitor or stored on disc. The images cover different sections of the globe and different spectral bands, scanned by Meteosat during the previous 25 minutes (see Chapter 13). An image is sent every 4 minutes, taking $3\frac{1}{2}$ minutes to build, followed by a half-minute gap before the next image starts to arrive. During this 30 second gap, the DCP data are retransmitted from Meteosat, all DCP data received at

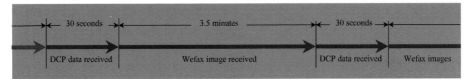

Figure 12.17. DCP data transmitted from field stations during the previous four minutes can be received via the Wefax channel in the half-minute gap between picture transmissions, which take 3.5 minutes to build, line-by-line.

the Meteosat ground station during the previous 4 minutes being disseminated (Fig. 12.17). A low-cost SDUS receiver is thus all that is required to obtain the data from DCPs in near-real time (with 4 minutes delay at most). The data rate is much faster than the 100 bits per second of the up-link used to send the data to the satellite. The data from all DCPs can be viewed, printed out or stored on disc, or just those with the required addresses can be selected.

However, these analogue transmissions are due to be phased out over the next few years, possibly by 2003, to be replaced by a new digital service called *low-rate information transmission* (LRIT) from the geostationary satellites (and *low-rate picture transmission* (LRPT) from the polar-orbiting satellites). The new service is due to start in 2001. Present receiving equipment can probably be updated (information on this and on costs can be obtained from Eumetsat's web page).

Cost of using Meteosat   There is no simple answer to the question whether the use of Meteosat to telemeter data has to be paid for, and if so how much it costs. It depends partly on the nature of the data: purely meteorological data of synoptic value are handled free if the data are coded for SYNOP transmission on the GTS. A charge may be made for other environmental data, such as hydrological or oceanographic data. If the project is sponsored by the WMO, the use of Meteosat will probably be free. It also depends which country is involved, those making a financial contribution to the European Space Agency (ESA) budget (the 17 member states of Eumetsat) being more likely to get free use. If a charge is made, it will be in the order of £1000 per year, per field station making three-hourly transmissions. This amounts to 2920 transmissions a year, making the cost of each transmission about 34 pence (50 cents). But charges change with time and Eumetsat should be contacted.

GOES

The geostationary operational environmental satellite (GOES) is the American equivalent to Meteosat and is operated by the National Oceanographic

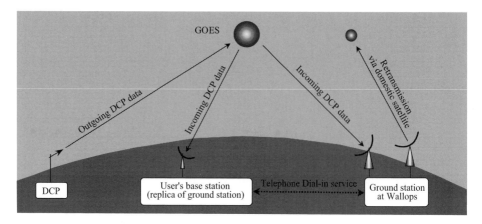

Figure 12.18. DCP transmissions can be received via three routes from the GOES satellites. In the US, they can be received directly by retransmission via a domestic satellite. Elsewhere there is a telephone dial-in service to the ground station. For larger organisations, it is possible to instal a replica of the ground-station receiving equipment at Wallops, although this is an expensive option.

and Atmospheric Administration's (NOAA's) national environmental satellite, data and information service (NESDIS), the ground station being at Wallops in Virginia. There are in fact two satellites (Fig. 12.8), GOES East and GOES West, covering half the earth's surface including all the Pacific region. Both collect images and support DCP telemetry, just as does Meteosat – but not in quite the same way.

Cost of use   An important difference between GOES and the other three geostationary satellites is that it is not intended exclusively for NWS use and weather forecasting. As its name indicates it has a broader role, encompassing environmental applications generally, and can normally be used free of charge for telemetering most types of environmental data. However, the three-hourly synoptic time-slot seems to be adhered to. Any country within the range of the two GOES satellites can use them.

Routes taken by data   While data suitably encoded in the SYNOP format are distributed over the GTS, there is no retransmission of DCP data from the GOES satellites on the Wefax channel, (although there is a Wefax image transmission service). The alternative is the GOES dial-in service, which allows users (in any country) to download their data by telephone by calling a number at the Wallops Command and Data Acquisition Station. The only cost of using the GOES satellites is the cost of the telephone calls.

For users outside the USA who are operating a large network of DCPs, it is possible to instal a replica of the NOAA ground station and thereby to receive the DCP data directly from the satellite. However, this type of receiver is very much more expensive and complex than one of the SDUSs used to receive Wefax signals, and requires a five metre receiver dish. It is not, therefore, an option that many users can afford, and service requirements also need to be considered. These various options are shown in Fig. 12.18.

GMS

The *geostationary meteorological satellite* (GMS) is operated by the Japan Meteorological Agency (JMA), and, as its name spells out, it is intended for synoptic and weather forecasting use. Indeed, other environmental applications are not allowed unless the DCP also takes readings appropriate to forecasting, such as barometric pressure or temperature. Nor is there any retransmission system or dial-in service, data having to be collected via the GTS. In addition to the organisational difficulties this can introduce, it also means that the data have to be formatted in the SYNOP or similar code. GMS is thus very much geared to meteorology, and other environmental applications are difficult to arrange. It would thus be useful if there was an alternative satellite in the region. The Russian GOMS satellite could be such an alternative but here, too, there are problems at present.

GOMS

The Russian *geostationary operational meteorological satellite* (GOMS) had been on the point of being launched for years and, as the name again indicates, is meteorological in aim. It was finally launched in 1996 but is not yet fully commissioned, with problems that need resolving before it becomes operational. However, it would seem that DCP data may only be accessible over the GTS, as with the GMS satellite, which again would limit its usefulness to broader environmental applications. Only time will tell, however, exactly what the satellite can offer, but it does seem that with GMS and possibly GOMS not offering a retransmission facility nor a dial-up service for DCP data, that this half of the globe is poorly served for environmental telemetry by the geostationary weather satellites, although well served for meteorology.

Indeed, with Meteosat's restrictions over non-meteorological use (from the payment point of view), only the area covered by the two GOES satellites is well served, and even here it is necessary to collect DCP data by telephone, which could prove expensive. For the synoptic meteorologist there is no problem with any of this, but something needs to be done to open up the use of all these satellites for hydrology, oceanography and other environmental sciences. Or perhaps the Inmarsat system, or the Argos system of polar-orbiting satellites, is the solution.

Data could now be distributed via the Internet by e-mail at very low cost and this would appear to be starting. However, it depends entirely on whether satellite operators are prepared to provide this option. It would certainly help.

Inmarsat

The international marine satellite system (Inmarsat) started operation in 1982. It is a commercially operated communications network based on four geostationary satellites (Fig. 12.8) giving worldwide coverage (except in the high-latitude polar regions, which are out of range of all geostationary satellites). It was initially intended for marine communication, and this is still one of its more important applications, but it now supports the transmission of voice, fax, video and data for a wide variety of uses, including the telemetry of environmental data. Inmarsat is operated in several modes.

Inmarsat-A is the original service (1982) offering direct-dial telephone, fax and e-mail. It requires a dish antenna and is full duplex. Thousands of ships are equipped with it. Inmarsat-B was introduced in 1993 and is the same as *A* but uses digital techniques rather than the analogue mode of *A*, allowing more efficient use of the satellite and so a reduction in cost. *Inmarsat-E* is for global maritime distress alerts. Inmarsat-M is a more compact, portable system, used, for example, by reporters in remote areas. The sound quality is lower than that of A, more akin to telephone quality. It can also be used for low-speed fax transmission. Inmarsat-D is the latest introduction, using pocket-sized terminals to transmit messages of up to 128 characters. Inmarsat-C is a compact system used for the transmission of low-speed digital data or text messages. This is done on a two-way *store-and-forward* basis, not in real time. (Inmarsat A, B and M offer full-duplex communication.) Inmarsat-C is the mode most suited to environmental telemetry and is our only concern here, although M and D might later find a place in such applications.

Routes taken by data   Figure 12.19 illustrates the route taken by data from a field station to and from the base station using Inmarsat-C. The data rate is 600 bits per second. There are two ways of operating such a system: self-timed or polled. In the former, field stations send their data at fixed times to the satellite, which retransmits them to one of its *land earth stations* (LESs). There are many LESs around the world – at a recent count, about 36. The received

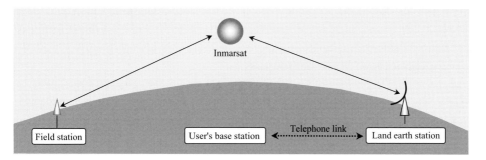

Figure 12.19. Unlike the weather satellites, the Inmarsat satellites are duplex in operation, allowing two-way communication. The field station transmits its measurements via the satellite to a land earth station (LES), where they are stored and then forwarded to, or collected by, the user's base station, by telephone. The base station can also send information to the field station by a similar but reverse route. Although not illustrated here, the base station can also communicate with the LES via the satellite, using an Inmarsat field station transmitter–receiver unit in place of a telephone link.

message can then either be held in an e-mail-like box at the LES until the base station collects it by telephone, much as in the case of the GOES dial-in service, or it can be forwarded to the base station as soon as it is received, provided that the base station is in readiness to receive it. Field stations can send their data as often or as infrequently as wished, there being no need to fit in with three-hourly transmissions as there is with the geostationary weather satellites, and therein lies one of Inmarsat's advantages.

Messages can also be sent to the field station from base via the LES through a reverse route, and so a polling network can be operated. But there is a problem with this, for, in order to receive a call from base, the field station receiver must be switched on all the time, and this may take more power than is available at many field stations. To get around this difficulty, it is possible to programme the field stations to call the LES at preset intervals, say daily, to collect any new instructions sent to the box number from base. The receiver then has to be powered up only for a time long enough to receive the new instructions, which might concern such matters as the time interval at which they should report. While this is not as versatile as being able to poll field stations in real time, it does give considerable more flexibility than is provided by fixed three-hourly transmissions to the weather satellites. This compromise can also be used with UHF ground-based telemetering field stations, which must also conserve power for the same reasons.

Although the Inmarsat satellites are geostationary, and so stay at a fixed point in the sky, Inmarsat terminals are normally mobile (for use on trucks and ships) and so an omnidirectional antenna is used. The antenna is small and unobtrusive, and so is less subject to vandalism than the much larger, higher-profile DCP Yagi antennas. However, it is possible to use a helical antenna, directed at the satellite, reducing the transmitter power required and thus the power consumption of the field station.

Cost   The cost of communicating via an LES varies from country to country, and so it is not possible to give a general estimate. The service provider in each country must be contacted. The cost also depends on the amount of data sent. If this can be limited to 32 bytes, the cost can be kept low.

Argos

The Argos system differs from the other satellites, being in a polar orbit. The Russian Meteor system is similar to it. Argos is run on a commercial basis, by the French space agency (CNES) through the company CLS (Collecte Local-isation Satellites) based in Toulouse, in co-operation with NASA using the NOAA satellites.

Method of operation   Currently there are two Argos satellites in general operational use (although a third is in orbit and can be used if necessary) and, with orbital times of about 100 minutes, each satellite passes over the poles 14 times a day, giving a total of 28 passes. The total number of passes by the two satellites combined, at any point near the equator, is six to eight. Because the satellites are not visible all the time, transmissions from field stations cannot be made at fixed intervals, but instead are repeated continuously every 100 to 200 seconds. When a satellite comes into view it will receive one or several transmissions, depending on the exact path it takes relative to the field station, before passing beyond the horizon again (it may not pass immediately over-head).

   Because the satellite is moving, an omnidirectional transmitter must be used. This means that the transmitted power cannot be concentrated by a Yagi antenna towards the satellite. However, because the satellite is in a low orbit and passes close to the field stations (compared with the distance to a geos-tationary satellite), less radiated power is required and this is why an omni-directional antenna is practicable. Because of the low orbit, the satellite is only in radio contact up to a distance of about 2500 km. Twice this range has been reported, and may at times be feasible, but a distance greater than 2500 km cannot be wholly relied upon in planning a network. Field stations in valleys, for example, will have a more restricted view of the satellite (this is also true for geostationary satellites), which reduces the range of communication and also

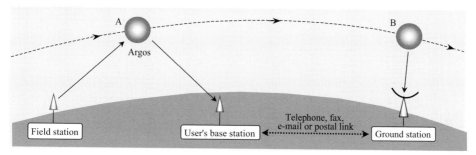

Figure 12.20. Argos field stations transmit their data every 100 to 200 seconds (on a frequency of around 401 MHz). When the satellite passes into view, the data are received and immediately retransmitted (on about 136 MHz). This retransmission can be received by a base station (provided that both the field and base stations are in view of the satellite at the same time – position A). This enables direct reception of data up to a radius of about 2500 km centred on the base station. The transmissions are also stored on the satellite and retransmitted to one of the three ground stations in the US and France (position B) as the satellite passes over. This provides an alternative to bases that are out of range of direct reception or in situations when real-time data are not required.

the time during which communication is possible; in flat terrain, the satellite is in view for a maximum of about 10 minutes at each pass, but in valleys it could be less. Since the data are transmitted every 100 to 200 seconds, they will be received several times during each pass. But because of this rapid repetition rate, the message length is limited to 64 bytes, which may not be enough for some applications, especially near the equator where passes are less frequent.

Routes taken by data   Figure 12.20 shows the path taken by data via Argos. Upon reception at the satellite, the message in processed in two ways: it is immediately retransmitted and is also stored on board the spacecraft. With suitable equipment, users can receive the retransmissions directly in real time, with no delay, provided that both the field station and the receiver have a view of the satellite at the time of transmission. If not, then the data must be collected later, after the satellite has passed over one of the Argos *ground stations* – one being in France and two in the USA. As it passes over a ground station, the satellite downloads all of the data collected since its last overpass and these are either disseminated on the GTS, sent by post, or collected by telephone via a dial-in service similar to that for GOES. But with ground stations only in France and the USA, telephone calls could be expensive for some countries, although e-mail is now a cheaper alternative being made available.

Avoidance of interference between DCPs   To reduce the probability that two field stations will transmit at the same time and thus interfere with each other, slightly different repetition intervals are allocated to each field station in any one area (in the range from 100 to 200 seconds). Should, despite this, two stations transmit at the same time, there will be a subsequent second or third opportunity to send the data while the satellite is still in view, since the stations, having different repeat times, will not interfere again for some time. If there are many stations then the chance of simultaneous transmissions increases, but there is a second strategy to overcome interference, as follows.

Position location   The Doppler shift of the received radio signals that occurs because the satellite is moving relative to the field stations results in each DCP's transmissions being received at slightly different frequencies, since each has a different geographical position, and this is sometimes sufficient, on its own, to separate two simultaneous transmissions. The shift in frequency has an additional benefit, for it can locate the position of field stations, so that not only can data be collected from the field stations but the station's location can also be determined and transmitted. This is one of the strengths of the Argos system where field stations are mobile, such as drifting buoys at sea (Chapter 13) and in animal experiments. No other satellite with a DCP relay facility has this capability, although with GPS now available, which can be used as one of the inputs to a DCP, this advantage is lessened. Nevertheless, Argos retains the advantage of not requiring the addition of a GPS receiver.

Cost   All use is chargeable and, while again it is difficult to pinpoint costs exactly, they can be anywhere up to 26 000 French francs per year per field station, although this does vary.

Polar versus equatorial cover   Argos and Meteor are the only weather satellites that cover the polar and high-latitude regions, at present (although Eumetsat is to introduce three polar satellites in 2003). Indeed, they cover these high latitudes better than the equator. But therein also lies one of their disadvantages for, at the equator, passes are limited to six or seven times a day, with occasional gaps of up to $6\frac{1}{2}$ hours. But in those large regions of the world covered by the GMS and GOMS satellites, which lack a retransmission facility or a dial-in service and which require the dissemination of data in the SYNOP format over the GTS, Argos may be seen as the better choice, even if it has to be paid for, despite its reduced time coverage at lower latitudes.

Orbcomm

Just going into service is a new network of satellites in low orbit known as Orbcomm, one satellite always being in range from anywhere on earth. They are not weather satellites but communication satellites intended to carry very short rapid bursts of digital data. There is already action to use this network for environmental telemetry, but it is too new to comment upon further at the moment. However, it would seem that the system will be most economical if used to send very small amounts of data (6 to 250, typically 20 bytes), not the larger volumes often sent through the weather satellites.

Which satellite is best suited for environmental telemetry will, therefore, depend on very many factors, all of which have to be weighed against each other. And as the situation is changing rapidly, with new satellites regularly appearing, the above is just a snapshot of the situation in late 1999.

References

Eumetsat (1995a) Meteorological data distribution, user guide. EUM UG 01.
Eumetsat (1995b) Data collection system, user guide. EUM UG 02.
Eumetsat (1996) The Meteosat system. EUM TD 05.
Eumetsat (1997) The Eumetsat polar system. EUM BR 06.
Eumetsat (1998) Meteosat high resolution and Wefax imagery. EUM UG 03.
Harris, L. (1996) *Satellite Projects Handbook*. Newnes (Reed Elsevier), ISBN 0 7506 2406 X.
Schafer, G. L. & Verner, J. G. (1996) SMOTEL, into the year 2000. in *Proc. Am. Met. Soc. 12th Conf. on Biometeorology and Aerobiology, January 1996, Atlanta*.
Strangeways, I. C. (1985) Automatic weather station and river level measurements telemetered by data collection platform via Meteosat. In *Proc. Int. Workshop on Hydrologic Applications of Space Technology, Cocoa Beach*, pp. 194–204.
Strangeways, I. C. (1990) The telemetry of hydrological data by satellite. Institute of Hydrology Report No. 112, p. 60, ISBN 0 948540 222.
Strangeways, I. C. (1994) Satellite transmission of water resources data, Technical Report 42. In *Proc. WMO Regional Workshop on Advances in Water Quality Monitoring, Vienna*, pp. 292–301. WMO/TD No. 612.
Strangeways, I. C. (1998) Transmission of hydrometric data by satellite. In *Hydrometry: Principles and Practice*, second edition, ed. Herschy, R. John Wiley & Sons, pp. 245–64. ISBN 0 471 97350 5.
Strangeways, I. C. & Lisoni, L. (1973) Long-distance telemetry of data for flood forecasting. *UNESCO, Nature Resources*, **IX**, 18–21.
WMO (1988) Manual on codes, Volumes I and II. WMO No. 306.

13

Oceans and polar regions

The oceans cover 71% of the earth's surface and cold regions occupy all latitudes north and south above about 65°; if we add to the oceans and cold regions the mountainous parts of the world then, together, these environments account for a major part of the globe's surface and yet it is difficult to make measurements in any of them. Deserts and forest, which are either sparsely populated or inhospitable or both, further reduce those parts of the planet that have been measured adequately. This chapter investigates what the problems are and what can be done about them.

The oceans

Sensor compatibility

All the sensors covered under the topic of fresh water measurement (Chapter 10) can also be used in the sea. These include water level staff gauges, float and pressure water level sensors for tidal records (Fig. 13.1), all the water quality sensors (Warner 1972) and current meters. The same sensors that measure atmospheric variables over the land are equally suitable for their measurement over the sea, although there may be problems with how to expose them. Also applicable to use at sea without any modification are data loggers and telemetry systems. So rather than there being a completely different range of instruments for the ocean, it is more a matter of how the instruments are deployed that differs between land and sea applications. There are two main ocean platforms on which instruments are operated – ships and buoys (to which might be added small islands).

Oceanography is researched by just a few major institutes around the world, from which workers ply the seas on short or long missions to make specialised observations for basic research – rather, that is, than making routine observa-

Figure 13.1. In most cases, the same instruments used for measuring fresh
water can be used to measure the sea, the example shown here being one of the
earliest sea level monitoring stations in Antarctica, at the BAS base at
Faraday; a conventional float-operated chart recorder is operational in the
hut over the stilling well.

tions. An example of this is a study by the Southampton Oceanography Centre
in the UK of wind stress at the sea surface, based on the *inertial dissipation*
technique (Yelland *et al.* 1994), which uses sonic anemometers mounted on a
ship. For this type of study, arrays of sensors may be fixed to the ship's
superstructure and connected to an on-board PC or data logger, or buoys may
be used. This is specialised research, but it is important to recognise that the
equipment used in most of these investigations is the same as that already
described for use on land.

Ships as instrument platforms

A short history of meteorological measurements from ships   From the voyages
of the seventeenth century onwards, meteorological (along with astronomical,
cartographic, geomagnetic and oceanographic) observations have been made
at sea (Kenworthy & Walker 1997). But it was not until the first International
Meteorological Conference was held in Brussels in 1853, at which ten maritime
nations met, that the scientific and commercial value of weather observations
from ships was fully recognised. The conference was largely at the instigation
of Lieutenant Matthew Fontaine Maury of the US Navy. As a direct result of
this conference, the British Meteorological Office was established as a depart-

ment of the Board of Trade in the UK in 1854. It was first headed by Captain (later Vice-Admiral) Robert FitzRoy (Barlow 1994, 1997), earlier Captain of the *Beagle* which took Charles Darwin on his voyage of discovery. By 1855 FitzRoy had obtained the co-operation of 105 merchant ships and 32 ships of the Royal Navy in making weather observations at sea, equipping them with the necessary instruments. Measurements were initially recorded in a *weather register* agreed at the Brussels conference. This was improved in 1874 by the Marine Superintendent of the Meteorological Office, Captain Henry Toynbee, who drew up the *meteorological log*, which was then recognised internationally. These logs remained unchanged for about 45 years, but after about 1921 they were gradually discontinued in favour of observations made at synoptic hours (00.00, 03.00, 06.00, etc.) and transmitted by radio. Until this time observations had been made six times a day, at the end of each watch. In 1953 and again in 1982 the layout was changed, resulting in today's *meteorological logbook*, which combines observations made and radio messages sent. Small changes were made in 1988, 1990 and 1994. Because of the importance of maritime meteorology, the International Meteorological Organisation was set up in 1872; this later became the World Meteorological Organisation (WMO), in 1951.

All observations from ships today are made on a voluntary basis, so as to get the required dedication and high quality required – it is better to have no data than poor data. The ships are known as the *Voluntary Observing Fleet* (VOF) and the observers as the *Corps of Voluntary Marine Observers* (VMO). Each NWS recruits its own ships, and there are regular visits to the ships by foreign NWS to supply materials and attend to instruments. For a time there were also dedicated *ocean weather ships* (OWSs), the sole purpose of which was to observe and report the weather from fixed locations, such as in the mid-Atlantic. But cost has meant that, in the North Atlantic, numbers have fallen from eight in the 1970s to just one today (a Norwegian ship).

There are three classes of ship today in the VOF. *Selected ships* observe wind, 'weather' (a summary of *present* and *past weather* in a specially coded form), barometric pressure (and its tendency), air and sea temperature, humidity, clouds and waves. *Supplementary ships* make the same observations less pressure-tendency, sea temperature and waves. *Auxiliary ships* are similar to supplementary ships but do not report cloud (under 'weather', see above). The *Marine Observer's Handbook* (Met. Office 1995) describes all these in detail, as well as the observation of sea-ice and other marine and atmospheric phenomena.

Of these observations only barometric pressure, air temperature, humidity and sea surface temperature are actually measured by instruments. The rest are non-instrumental observations and are, therefore, to some extent beyond the terms of reference of this book, but as they overlap instrumental observations, something will be said about them.

Barometric pressure   There is evidence that Robert Hooke, in 1667, was the first to see the advantage for a ship of having a barometer, and he presented a paper to the Royal Society on his experiments to overcome the problem of oscillations of the mercury due to the ship's motion (Banfield 1976). Edmund Halley took such a barometer on one of his voyages of exploration in the Atlantic in 1698. One of Captain FitzRoy's initial tasks after taking up his position as head of the Meteorological Office in 1854 was to supply ships with mercury barometers. He also designed a barometer that would resist the harsh conditions on-board a sailing ship, including the damage that the firing of a ship's guns could do (Barlow 1994). The present-day Kew-pattern marine barometer (Chapter 7), a derivation of the FitzRoy design, reduces the effects of 'pumping' of the mercury due to wind and movement of the ship by making part of the tube in the form of a fine capillary, opening out to full bore only towards the top, to allow the meniscus to be read easily. This design remained virtually unchanged since it was first made in about 1900, but it has been replaced by aneroid barometers, which can now be made having a similar accuracy to the Kew marine instrument while being more compact and easier to read. As on land, on-board a ship a barometer is installed in an office, since it requires the same precautions against wind effects.

Air temperature and humidity   Temperature and humidity are measured in the same way on a ship as on land, but using a smaller Stevenson screen, just large enough to hold the wet-and-dry thermometers and nothing more. No chart recorders or maximum and minimum thermometers are used, just spot observations at the specific locations at the time. The screen must be exposed, so as to be well away from any heat sources such as air from the boilers or the living quarters, the rails on the bridge being a typical choice. However, there cannot be the same consistency of exposure as is possible on land, since ships differ in size and structure. Also, while the wind will sometimes be directly off the sea, at others it will blow across the warmed deck. And, on larger vessels, the temperature tens of metres above the sea surface will be measured and this is bound to introduce some uncertainty. The same type of mercury thermometers used at land stations are suitable at sea, as are electrical resistance thermometers if a remote display is needed. In the nineteenth century, most thermometers on-board ships were not screened and so there is doubt as to the reliability of readings from the days of sail (Chenoweth 1996).

Measuring sea-surface temperature manually   Because of the influence that
the sea has on the world's weather and climate, sea-surface temperature is of
great interest to both meteorologists and climatologists. This has been measured since marine observations began in the nineteenth century, and there are
several ways of making the measurement.

Manual measurements are made by collecting a sample of surface water in a
bucket on a rope. Upon withdrawal the thermometer is inserted immediately
into the water sample, away from sunlight, and submerged up to the top of the
stem, stirring all the time, being withdrawn after about a minute just far
enough to take the reading. The observer's hand must not warm the water. A
thermometer of the type used to measure air temperature is suitable. Buckets
vary in design, but a typical one is made of double-skinned canvas or rubber.
Single-skinned buckets are not suitable since evaporation on the outside
surface can cause cooling. As the speed of the ship or its height increase, the use
of a canvas bucket becomes increasingly difficult and the UK Meteorological
Office now supplies a heavier rubber model. As the ship's speed increases, care
is also needed to prevent even the heavier buckets from skimming along the
surface, collecting a sample just from the very top layer or even of sea spray,
neither of which will be at a representative temperature. To avoid this, by
ensuring adequate mixing, the bucket must submerge 'cleanly', ideally to a
depth of about one metre.

Remote-reading sea-surface temperature thermometers   There is an advantage
in reading the thermometer while it is immersed in the sea, and although it is
possible to lower an electrical resistance thermometer on a cable from the
bridge into the sea, it is difficult to control the depth it takes up. Instead, it is
more usual to fix a platinum resistance thermometer to the inside of the hull, a
metre or so below the normal water line, the conduction of the steel being good
enough for it to take up the temperature of the outside water. As the hull has a
large mass, the mean temperature is sensed and this compensates for any
rolling or pitching of the ship.

Engine-room intake temperatures   If hull temperatures cannot be read, an
alternative is to measure the water as it is pumped into the ship via the intake
to the engine-room. However, as the ship rolls and pitches water will be drawn
in from different depths and this can cause some error. Also there can be
heating if the temperature is not measured very close to the intake, because the
water passes along pipes within the warm ship. Much work has been done
recently to assess the accuracy of sea-surface-temperature data, with regard to
climatic change (Folland & Parker 1995, Chenoweth 1996, Parker *et al.* 1995),

and to assess the quality of data collected by the various methods over the last 150 years.

Measuring temperature profiles   An instrument known as a *bathythermograph* is used to measure the change of temperature with depth. In the past this took the mechanical form of a bimetallic strip (of the type used for measuring air temperature) that recorded a trace on a smoked-glass slide, the whole being contained in a bronze, torpedo-shaped housing and lowered and recovered by winch. After recovery, the slide was read against a graduated scale. However, today's alternative, used on weather ships, research vessels and ships of opportunity, senses the temperature with an electrical resistance thermometer, its measurements being recorded on a data logger as its cable is lowered. Profiles of temperature, however obtained, give an indication of subsurface currents, thermoclines and long-term climatic change.

Wind speed and direction   Although anemometers and wind vanes of the type used on land operate perfectly well on ships, there are two problems. The ship's forward motion, and its rolling and pitching, affect the readings and correction must be made for this, but more problematic still is finding a suitable place to fix the sensors, since the ship disturbs the airflow. On research vessels (Birch & Pascal 1987) duplicated sets of sensors can be sited at several places around the ship to ensure a fair exposure of at least one set at any one time. But this is not practical for the ships of the VOF, and for these the preferred method is to estimate wind force and direction by observing the sea-state. Even though this is not a measurement but an observation, it is worth saying something about it.

Wind force expressed on a scale of 0 to 12 was originally devised by Admiral Sir Francis Beaufort in 1806, based on the sail canvas carried by a frigate of the Royal Navy (Chapter 5). As ships changed the scale was adapted, but, with the passing of sail, the practice arose of judging wind force by the appearance of the sea corresponding to each Beaufort number; this new scale came into use in 1941 (Met. Office 1995). This is the method currently used by the VOF. However, judgement and experience are necessary since the state of the sea is also dependent on such additional factors as the closeness of the shore, the state of the swell (which results from winds a long way from the point of observation or from more local winds now ceased) as well as tides and currents. Also, the difference in temperature between the air and the water affects the waveform, while rainfall can have a smoothing effect on the sea surface. How long the local wind has been blowing also affects the sea state; it takes time for waves to build up. The quality of the observations is thus very observer-dependent and is nothing more than an estimate.

Rainfall   Many factors reduce the effectiveness of a raingauge when operated on a ship rather than on land, and even on land its exposure is difficult enough. Rolling and pitching will require that the gauge is kept horizontal by some form of gimbal – not just to ensure reliable tipping bucket operation, but also to collect the correct amount of rain initially. But the higher winds experienced at sea, together with the forward motion of the ship and the very open exposure, make the wind-induced errors even greater than those experienced on land, further worsened since the gauge has to be exposed amongst the ship's superstructure. To these problems must be added the possibility that sea spray will be collected along with the rain. There are no simple answers to these problems, the best probably being the use of aerodynamic collectors (Hasse *et al.* 1997, Folland 1988, Strangeways 1996). Yet a further problem is that the rainfall total will refer not to one geographical location but to the whole area across which the ship has travelled during the collection period.

Observing ocean waves   The study of ocean waves was only recently put on a scientific footing, with the development of automatic wave recorders producing quantitative measurements. These have shown up the limitations of the previous observational methods. Ocean waves are complex and impossible to analyse simply by inspection, but when electronically recorded they can be analysed into the simple component-waves that combine to form them. But wave recorders can only be effectively used on stationary installations such as oil rigs, (stationary) weather-research or oceanographic-research ships, or buoys. For the VOF, wave recorders are not, therefore, an option and the old observational procedures still have to be used.

Apart from the few waves caused by tidal effects, all waves are due to wind – although how the wind produces them is still not precisely understood. Waves raised by the local winds blowing at the time of the observation are referred to as 'sea'. Those due to local winds now ceased or to winds a large distance away are known as 'swell' and are long in wavelength (hundreds of metres) compared with locally generated waves. Although there can be swell from two directions at once it is usually just from one, but this will be combined with the 'sea', which as a rule will be from a different direction.

It is difficult to observe this complex motion of the surface, but certain rules have been developed. Three characteristics have to be noted in a ship's report. The *direction* from which the waves come is relatively easy to estimate by sighting either across the wavefront or along the crests. The *period* is measured with a watch by selecting a patch of foam or small floating object and measuring the time between the larger waves (of each successive wave group), the process being repeated 20 or more times to get an average. The *height* is not so easily estimated and there is no suitable method for general use on merchant ships. However, if the wavelength is less than the ship's length, the height can be estimated roughly by the appearance of the wave on the side of the ship, while if it is longer than the ship the observer can take up a position such that the top of the wave is level with the horizon. The height of the wave is then equal to the height of the observer's eyes above the ship's water line.

Estimating ocean currents Knowledge of ocean (surface) currents has been acquired over 100 years or so, mostly through observations from ships on-passage. Much remains to be discovered, however, especially in areas remote from the main shipping lanes and about the variations in currents over time.

Currents are measured from a moving ship by estimation of its position using dead-reckoning (that is, calculating its position by measuring its speed, direction and transit time between two points), with correction for leeway-drift due to winds, and comparing this with the actual position, today more easily known from Global Positioning System (GPS) instruments. The vector difference between the two positions is that caused by currents. The method is assumed to give the current at a depth of about half the draft of the ship, but is not used over distances of more than 400 miles, or travelling times of more than a day, as several different currents may then be involved. But too short a time is also undesirable because of the small differences involved. This method is not very precise, and it is also liable to error particularly because it is difficult to estimate the drift due to wind precisely enough. We are still very short of good ocean-current measurements.

Upper-air soundings – radiosondes An important rôle played by ships is the launching of radiosondes to measure the atmosphere up to around 3 km height. Radiosondes are lightweight expendable packages, which are taken aloft by helium-filled balloons, about 1 metre in diameter, carrying sensors to measure air temperature, humidity and barometric pressure. Measurements are transmitted to the ground-launching station on a frequency of 401 to 406 MHz, as a continuous sequence of repeating patterns of data, including some reference channels, typically at a repeat time of about 1 second per variable. These are received at the ground station and automatically converted into readings that can be formatted into SYNOP for transmission by a data collection platform (DCP) via satellite onto the GTS.

In one recent design of radiosonde, temperature is measured by a newly developed very small sensor made of two fine platinum wires encapsulated, parallel to each other, in a glass ceramic material that acts as a variable dielectric, changing the capacitance between the wires as the temperature changes. The range covered is from $+55$ to $-90\,^\circ$C. Because there is no temperature screen on a sonde (the sensor being exposed directly to solar radiation) a very small sensor of this type is needed, in order to minimise radiative heating.

The humidity sensor is also capacitive, and has been described already in Chapter 4. It differs from the ground-based sensor, however, in that it has an integral heater on the chip to combat wetting in cloud and ice formation on its surface. Two sensors are used, heated in turn, the currently non-heated sensor being used to take the measurements. These precautions are necessary because early sensors were badly affected by passage through cloud. (Exactly what the relative humidity (RH) is in a cloud is debatable, for the cloud may be supersaturated, in which case the RH would be over 100%.)

Barometric pressure is measured over the range 3–1070 hPa, either by a miniature sensor in which a silicon diaphragm flexes under a change in pressure, causing a change in capacitance, or by an aneroid capsule producing a similar capacitive signal (Chapter 4).

Wind speed and direction can be measured in several ways, currently the most usual being to track the sonde by radar, the speed and direction of the wind being deduced from this. But, increasingly, sondes find their location themselves, by using either the ground-based navigational system LORAN (*long-range navigation*), or the satellite based GPS system (by including additional electronics in the sonde), the position being transmitted to ground. Here it is compared with the position of the ground station, also measured using LORAN or GPS, the relative movement due to wind being deduced. There is a further wind-measuring system in which the sonde transmits its data on 1680 MHz, an interferometer-antenna at the ground station measuring the azimuth and elevation angle of the sonde, the height of the sonde being inferred from the pressure sensor. (In systems using radar, height is determined by trigonometry.) Sondes are launched from both ships and ground stations on a daily basis.

Moored buoys as instrument platforms

The UK situation will again be used as a basis for a more general consideration of the worldwide position. The UK Meteorological Office operates three types of buoy – moored buoys in deep water, moored buoys in shallow coastal waters and drifting buoys, these taking two forms.

Currently there are seven deep-water moored buoys on the edge of the European continental shelf west of the UK and Ireland, in water depths from 2000 to 3500 metres, including one off the coast of Brittany, with French co-operation. There are also two in the North Sea. The USA and Canada operate similar buoys off their Atlantic and Pacific coasts and across the Pacific, America currently having about 65 such buoys. At present there are four shallow-water moored buoys in the Irish sea and the English Channel. In addition, lightships are used as moored platforms, of which currently four are equipped with sensors; all these are in the English Channel. Use is also made of small islands and oil rigs to deploy similar instruments, and around the UK coasts there are at present seven, to the east and north.

Figure 13.2 illustrates a deep-water moored buoy of the UK design being visited by the *Salmaid*, a ship used by the UK Meteorological Office to instal and service this type of station. The hull of the buoy is 3 metres in diameter with a sensor ring 4.5 metres above sea level (Met. Office 1997). The variables measured are barometric pressure, air temperature and humidity, sea-surface temperature, wind speed and direction and wave height and period. With the exception of the wave sensor, all sensors are duplicated in case there are failures, as service visits are expensive and so limited to a set schedule.

Barometric pressure   A ceramic diaphragm sensor, vibrating cylinder or aneroid capsule (all described in Chapter 7) is used to produce a ten-second average reading of pressure at the synoptic hours. The sensor is housed on the outside of the buoy's superstructure, protected from the effects of wind by a *static pressure head* (Chapter 7). While some countries deploy their sensors inside the buoy's superstructure whenever possible, the UK policy is to deploy them externally to simplify interchange.

Air temperature and humidity   A platinum resistance thermometer (PRT), housed in a miniature screen of the AWS type (Fig. 3.8), measures air temperature to $\pm 0.2\,°C$ with a resolution of $\pm 0.1\,°C$.

The humidity sensor is housed in a separate but identical screen, to allow its exchange independently of the temperature sensors. Unlike most land-based AWSs, the buoys use an ion-exchange humidity sensor (Fig. 4.6). Along with all sensors except the sea-surface-temperature and wave sensors, the humidity sensor is changed every six months (weather permitting). Experience has shown that after nine to twelve months the calibration of the humidity sensor starts to drift, probably due to the salty environment; they are not reused. The sensor is protected from direct contact with the atmosphere and sea-spray in a cavity covered by a PTFE membrane (Clarke & Painting 1983), which is permeable to water vapour but not to liquid water. This type of sensor is preferred to the capacitive type because the humidity is usually high over the sea (75% to 100%) and the resistive sensor works well in this range (having an

Figure 13.2. The UK Meteorological Office operates nine buoys of the type illustrated, moored along the edge of the European continental shelf; this one is receiving a six-monthly service visit. The wind sensors and temperature screens can be seen on the upper ring, with solar panels on three sides. Sea-surface-temperature sensors are fixed to the hull one metre below average sea level, and the barometric pressure sensors are protected within the super-structure. An inertial wave sensor is housed in the main central cavity. Measurements are telemetered via Meteosat. (Figure reproduced by permission of the UK Meteorological Office.)

accuracy of $\pm 5\%$ below 85% RH and of $\pm 3\%$ above 85% RH). When the RH measurement is combined with the temperature measurement, dew point can be calculated to $\pm 0.4\,^{\circ}\mathrm{C}$ with a resolution of $0.1\,^{\circ}\mathrm{C}$. The value measured is an instantaneous reading at the observing time.

Sea-surface temperature   As on ships, a PRT sensor, fixed to the hull of the buoy, measures the water temperature at one metre depth to $\pm 0.2\,^{\circ}\mathrm{C}$ with a resolution of $0.1\,^{\circ}\mathrm{C}$. Because it is possible to locate the sensor more precisely at the required depth than on a ship, the data are more truly representative of temperatures at one metre.

Wind speed and direction   Conventional cup anemometers and vanes are used (Chapter 5), although the vane is of the self-referencing type, with a built-in magnetic compass providing an automatic reference to magnetic north, making it unnecessary to determine the orientation of the buoy. The anemometers are oil-filled to protect the bearings from contamination by salt. In accordance with standard synoptic procedure, wind speed and direction readings are averages for the ten minutes preceding the observation time, made with an accuracy of $\pm 2\,\mathrm{kn}$ below $40\,\mathrm{kn}$ and $\pm 5\,\mathrm{kn}$ above $40\,\mathrm{kn}$, and with a resolution of $1\,\mathrm{kn}$. Gusts are also measured.

However, the exact height is debatable, since although the anemometer is deployed at four metres above the buoy, the buoy may be changing height by more than this amount because of the wave movement. In a trough the wind speed may be lessened, while the swaying of the tower introduces some further uncertainty. However, perforce, we have no alternative.

Wave measurements   It is not possible to use a pressure sensor to measure waves from a floating platform; it is necessary, instead, to use an inertial sensor, consisting of a passive accelerometer which measures the vertical movement of the buoy. Since the buoy has a considerable inertia, the smaller waves will not be detected as they will be averaged-out. An algorithm converts the detected direction and amplitude of movement into wave height and period. The *significant wave height* is defined as the rms (root-mean-square) value of the water level above the average level, measured over a period starting 17.5 minutes before the observing time; accuracy is $\pm 10\%$, or $\pm 20\,\mathrm{cm}$ if greater, over a range of $\pm 10\,\mathrm{m}$ relative to the mean level. The *wave period* is defined as the average time interval, during the 17.5 minutes, between successive passages through the mean water level in an upward direction; accuracy is to ± 0.5 seconds.

Rainfall and evaporation   Because rainfall is not required for forecasting purposes, it is not measured on buoys, the main purpose of which is to provide data for weather forecasts. It would, however, be a difficult variable to measure on a buoy, in part because of the unstable platform, which would make it necessary to support the raingauge on a gimbal to keep its collector and the tipping bucket mechanism level. In Chapter 8, two types of gauge that might be more suited to buoy operation in place of the tipping bucket were mentioned. In one, the collected water is measured capacitively. The other type is the optical raingauge, which is not as sensitive to levelling; however, it is very expensive. In addition, for any type of raingauge there is the possibility that sea spray will be collected and read as rainfall. Because of the problems of exposing a raingauge on a ship and because rain is not measured on buoys, our knowledge of rainfall over a very large part of the earth's surface is limited in the extreme.

Experiments are, however, being undertaken to measure rainfall at sea by the sound of the raindrops (Chapter 8), not the sound as they impact the surface, but the sound of the bubbles produced as a result. These sounds are analysed by an algorithm, the bubble sound-signals being distinguishable from amongst the myriad other noises of a stormy sea and even from the sounds of sea-spray bubbles, owing to their different drop-size spectra. However, since such techniques are not yet operational, it would seem worthwhile doing experiments using a tipping bucket raingauge mounted on gimbals, despite the undoubted problems that exist. The use of an aerodynamic rain collector (Fig. 8.6) would go some way towards minimising at least the wind effects, while having measurements of wind speed and wave activity would allow those periods when readings were probably correct to be determined automatically. Further, by measuring the conductivity of the collected water, contamination by sea spray could be detected and probably quantified (by knowing the dilution of the water). But this is not fact, just idle speculation.

Nor is evaporation from the oceans measured, since there are no calls for this to be done. While three of the four variables for estimating evaporation (Chapter 6) by Penman-type methods are already measured on the buoys, the fourth, radiation, is not. While it would not be a simple matter to measure net radiation (Chapter 2) on an unstable platform liable to sea spray, nevertheless solar radiation, or sunshine duration, probably could be measured with fair accuracy. Further, since the environment is open water, uncomplicated by the concerns on land of water availability and vegetation, the estimate should be more accurate. This lack of one sensor means that a second very important hydrological and climatic variable goes unmeasured over two thirds of the earth's surface. This highlights the problem that because NWSs are not asked to measure rainfall or evaporation over the oceans, there is no funding either to undertake the measurements or to develop the means to do so. It might be argued that since NWSs are not asked to make such measurements, there is clearly no need for them; but such data would be of undoubted value to climatology if available. This is an obvious gap in our knowledge.

It would be unrealistic to attempt to measure evaporation over the sea using the eddy correlation method (Chapter 6) on a routine basis, although it would be possible for short experimental periods. However, while interest grows in measuring the flux of carbon dioxide over forests (by a modified eddy correlation instrument), it is not yet being measured over the sea, surely earth's biggest carbon sink, reflecting the rather strict boundaries that still exist between the disciplines of meteorology, hydrology, oceanography and climatology, no doubt owing to the separate nature of their funding.

Logging and telemetry   Just as with any AWS in a remote location sending data in real time, a logger and the means to telemeter the measurements are required (Chapters 11 and 12); the moored buoys on the edge of the UK

continental shelf use Meteosat (Eumetsat 1995) to relay data to shore. The
logger and DCP, like the sensors, are duplicated, each DCP transmitting the
measurements from both sets of sensors in engineering units (not SYNOP).
Upon reception, the two sets of data are combined to give optimum results,
which are then coded in SYNOP for transmission on the GTS. As with
land-based AWSs, lead–acid gel batteries charged by solar panels power the
buoys.

Location systems   Although moored in deep water, buoys on the surface do
not move by more than a mile from their point of fixing and so need no method
of location. However, if one should break free from its mooring and drift, a
means to locate and retrieve it is necessary, owing to the high cost of replacement. Two methods are used, the first being the inclusion of a satellite GPS
receiver as one of the channels transmitted by the DCP, the second being an
Argos transmitter (Chapter 12). Two radar reflectors are fitted, allowing the
buoy to be detected by local shipping for safety reasons.

Inshore moored buoys   The moored buoys installed in shallower waters
(around the UK) are simpler than the deep-water type, although similar (Fig.
13.3). The sensors are the same, but, since the buoys are more easily visited,
there is no need to duplicate them. Similar stations are also operated on four
lightships in the English Channel and also on some islands and oil rigs.
Telemetry is by whichever means is most suitable, via either ground-based
UHF or DCPs (for example on oil rigs).

Drifting buoys as instrument platforms

By far the greatest number of buoys in use around the world are not moored,
but drifting, the greatest concentration of moored buoys being in the Atlantic
and Pacific oceans. There are two types of drifting buoy used in the Atlantic –
those that drift under the combined effects of wind and current (Fig. 13.4) and
those mostly driven by currents only, these being in the form of small floating
spheres (not illustrated) attached to a large drogue and acting as Lagrangian
tracers embedded in the moving fluid: they move with the currents and are
little affected by wind.

Launch programme and area of operation   Both types of buoy are deployed
from ships of convenience. In the case of the North Atlantic, ships sailing from
Iceland to the eastern coast of the USA, ships *en route* from Denmark to Cape
Farewell and ships from the UK to the Caribbean cover an area from the

Figure 13.3. The UK Meteorological Office also operates moored buoys in the shallower coastal waters around Britain, these being a scaled-down version of the larger, deep-water buoys, having just one set of sensors and telemetering by ground-based UHF radio links or by satellite, whichever is the most suitable. (Figure reproduced by permission of the UK Meteorological office.)

Canary Islands up to Iceland and from the European and African coasts to 45° west. The launch programme is co-ordinated by the European Group on Ocean Stations (EGOS 1996), set up in 1988 with participants from eight countries. EGOS maintains a continuous operational network of drifting buoys, numbering from 25 to 40, with numbers set to rise as the cost of the technology falls. They are not recoverable for reuse, but have a long lifespan, averaging 200 days. Their cost varies from about £3000 to £5000 for the basic model to £15 000 for the full complement of sensors. The paths taken by the buoys vary considerably, some moving a long distance and others circling in a small area, the paths being quite unpredictable.

Figure 13.4. Drifting buoys are by far the most common type, worldwide, there being many hundreds at any one time. They either drift under the combined influence of wind and currents or are largely submerged and fitted with drogues to move with the current only. The one illustrated is of the former type, on test at the UK Meteorological Office. The screen at the top of the buoy contains an air temperature sensor; a barometric pressure sensor is housed in the central column; and a sea-surface-temperature thermometer is fixed to the hull at one metre depth. (Figure reproduced by permission of the UK Meteorological Office.)

While there are few buoys of the moored type (beyond those operated by the UK and France in the North Atlantic, the North Sea and the Baltic, and those operated by Canada and the USA, many of the latter being in the Pacific), drifting buoys are quite commonly used worldwide. South Africa has 60 at present in the South Atlantic, while several countries, including the UK and France, operate drifting buoys in the Indian Ocean. Australia and New Zealand deploy their own buoys in the South Pacific and Indian Oceans (Eumetsat 1995). Because there is less land in the southern hemisphere to interfere with their movement, the life of these buoys is longer, often allowing them to circumnavigate the globe.

Variables measured   Both types of drifting buoy measure barometric pressure
(and its tendency) and sea-surface temperature. The larger type, as illustrated
in Fig. 13.4, also measures air temperature (the screen is seen on top of the
column). A few measure wind speed and direction. The temperature and
pressure sensors are the same as on the moored buoys. Although this is not
included on the buoy illustrated, wind speed may be measured by a Slevonius
rotor (aerofoils that rotate on a vertical axis in a way similar to cup anemometers) mounted beneath the screen. A vane, fixed to the central column of the
buoy (not on the example illustrated), causes the complete structure to align
itself with the wind, a compass-sensor measuring the bearing of the buoy and
thus of the wind. The holes in the vertical mast are ports to the barometric
sensor. Methods of measuring wind speed by detecting the sound signature of
sea waves in the band 8 to 16 KHz are being investigated.

Logging and telemetry   Because the buoys are moving, their measurements
are telemetered via the Argos satellites instead of Meteosat, since Argos can
also fix their position without the need to include a GPS receiver. However,
while the cost of using Meteosat is covered by the contributions made by the
individual participating countries of the member states of Eumetsat (Chapter
12), Argos has to be paid for, currently at 26 000 French francs per year per
buoy.

   The telemetered data can be received directly from the satellite at three *local
user stations* located in Oslo, Søndre Stømfjord and Toulouse. By using these
three receivers, instead of the usual one, the area over which buoy transmissions can be received directly from the satellite in real time is extended
(both transmitter and receiver must be in view of the satellite at the time of
transmission if the data are to be received directly). The primary method of
receiving the data, however, is by the store-and-forward option, via Argos
ground stations in the US and France (Chapter 12).

   Ice buoys are also operated in a similar way, allowing the movement of
pack-ice to be monitored, although of necessity they are deployed from the air
by parachute. This technique is also starting to be used for drifting ocean
buoys, since it is then possible to place them in the remoter areas not covered
by shipping.

Sea level monitoring stations

As part of the international *Tropical Ocean Global Atmosphere* (TOGA) project and the *Antarctic Circumpolar Current Levels from Altimetry and Island*

Measurements (ACCLAIM) project, measurements of sea level are made at a number of coastal sites worldwide; this is in order to study the interaction between air and sea, to provide ground-truth for comparison with satellite altimetry data and to obtain long-term measurements of sea level for global-climatic-change studies. At these sites, sensors fixed to the sea bed or to a harbour wall measure the pressure of the sea (Chapter 10) and its temperature, while on land barometric pressure is measured (Chapter 7). Readings are taken every 15 minutes, the logger being programmed to give hourly averages. Sea level is obtained by subtracting the air pressure reading from the sea pressure, whereas river level pressure sensors, which are vented to the atmosphere, require just one reading to be made. The processed data are stored in the logger as back-up and are telemetered every 12 hours via Meteosat, GOES or GMS, whichever is geographically appropriate. Each 12 hour transmission is of the last 24 hours' data, so that if one transmission is lost the data can be retrieved at the next transmission. The data are sent to the TOGA Sea Level Centre at the University of Hawaii and to the British Oceanographic Data Centre at the Proudman Laboratory, where they are made available to the international scientific community. Meteosat receives data telemetered from 12 sea level stations located in the Indian and Atlantic oceans (Eumetsat 1995).

Cold regions

The problems of low temperature

Low temperature in itself is not a major problem to instruments. Electronic and mechanical components can operate in temperatures down to $-40\,°C$ or lower, and batteries continue to function at these extremes, although perhaps with a reduced capacity. The main problems arise from the effects of snow and ice adhering to (or filling) the sensors, often accompanied by strong winds. The question considered here is how to protect sensors from their damaging and disabling effects. (Techniques for measuring the amount of snow falling, lying and melting are dealt with in Chapter 8.)

Two types of ice can form. *Hoar frost* is produced directly from water vapour in the air, but as the amount of vapour is small at sub-zero temperatures, this is not a serious problem and it lessens as temperatures fall. The most hazardous form of ice, *rime*, occurs when supercooled cloud droplets meet a surface and freeze, the ice building mostly into the wind, but also in the lee owing to eddies. Rime can be extremely thick and damaging (Fig. 13.5). *Glaze ice* forms when the cloud-drops have time to form a film before freezing. The amount and type of cloud-induced ice that forms depends on the wind speed, air temperature,

Figure 13.5. Rime ice can be a serious hazard to instruments; it is seen here encasing an AWS on Cairn Gorm in Scotland. One cup of the anemometer remains clear of ice (left); the net radiometer (lower right) is completely covered, as are the other, barely recognisable, sensors.

water droplet size and drop density. Even though the atmosphere holds less and less water vapour as the air becomes colder, rime can still be a problem down to the lowest temperatures of Antarctica. The worst rime ice is not, however, found in these conditions (as might have been imagined), but on mountains of mid-latitudes exposed to water-laden maritime winds, such as in the Scottish highlands.

Snow causes problems either when wet, by adhering to sensors and, perhaps then freezing, or as dry crystals, blowing like sand (Fig. 8.11) and able to enter the smallest of holes and completely fill cavities such as temperature screens.

De-icing manned stations

A brief history   Automatic and manual stations alike suffer from the snow and ice problem. There have been a number of manned observatories in cold places, a classic example being that operated on the summit of the Scottish mountain Ben Nevis, from 1883 to 1904 (Paton 1954, Begg 1984, McConnell 1988a,b). Its main purpose was not to study mountain climate but the vertical structure of the atmosphere in depressions and anticyclones (Roy 1983), similar observations being made at an observatory at Fort William at the foot of the mountain to provide data for comparison with those from the summit. Today, however, the mountain-climate aspects of the data are of equal interest.

Operators lived on the summit and made observations every hour during the day and the night. Chart-recording instruments could not operate in the temperature screen under the conditions that prevail there for much of the year, being in supercooled cloud for a high proportion of the time. On many occasions the screen became completely ice-covered or ice-filled and often the thermometers had to have ice removed from them before readings could be taken. Conditions sometimes prevented the operators from getting to the instruments at all. Because of the exposed conditions and high winds, precipitation measurements were also prone to high errors and anemometers could not be kept continually ice-free. Even with people present all the time, it was an impossible task to keep the instruments fully operational, and there were inevitable gaps in the records.

At about the same time as the Ben Nevis operation, a similar station was being operated in northern Siberia at the mouth of the Lena River (Schutze 1883). But these examples were unusual and relatively short-lived, although very important in amassing information never before collected. During World War II, Germany and Norway operated manned weather stations on Greenland, radioing back their measurements.

Modern Antarctic stations   The nearest modern equivalent to the pioneering installations in Scotland and Siberia are today's semi-automatic stations operated at bases in Antarctica. Typical are those operated by the British Antarctic Survey (BAS), which has several bases on the continent – at Rothera on the peninsula (and until recently also at Faraday) as well as at islands further north and at Halley on the mainland. At such bases, measurements are made by electronic or electrical sensors, of the types covered in earlier chapters; they are exposed in much the same way as manual instruments, with cables running some distance into the laboratories of the base. Here, mains power being available from diesel generators, the measurements can be processed by a PC, operating continuously and connected to a DCP, allowing the measurements to be processed into SYNOP for transmission over the GTS. In addition to inputs from the sensors, the stations accept manual inputs of non-instrumental observations such as cloud cover, state of precipitation, snow cover and similar variables, which are difficult or impossible to measure automatically at any location.

Ice and snow prevention and removal at these stations are entirely manual. A sharp hit with an ice axe is used to clear wind sensors, while accretions are

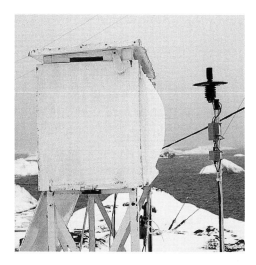

Figure 13.6. At Antarctic bases, the British Antarctic Survey prevents snow and rime entering or adhering to Stevenson screens by adding drop-down doors that can be closed when such hazards are occurring or are expected. Here the left-hand cover (at the back of the screen) is open and the other three are closed; the front is covered in adhering snow. Although this procedure reduces the air flow considerably, it is only necessary when radiation levels are fairly low, and so does not affect the temperature readings unduly; but clearly it is a compromise.

removed from radiation sensors by a gentle brush by hand. The Stevenson temperature screen is of conventional wooden construction, but has fold-up doors on all sides, which can be closed on the side(s) exposed to the wind when there is snow or the possibility of rime formation (Fig. 13.6). Rainfall is not measured and snow is monitored by snow poles.

Even though the stations are manned continuously, the manual de-icing methods obviously have their logistical problems and limitations, just as they did on Ben Nevis. No manual system can ensure continuous observations. If snow starts to fall, or rime to build up, it is necessary to go out, lift the appropriate door on the screen and climb the 10 metre high wind-sensor mast (Fig. 13.7) in order to remove the accretions manually; this can be difficult or impossible in high winds or at night, if for no other reason than safety. Working in this way, the sensors (temperature apart perhaps) cannot be kept continuously ice-free and although for much of the year these methods are adequate or are not required at all, inevitably there will be periods when the data are not correct. However, it will generally be clear when the data are suspect and these can be flagged as doubtful. Users of such data should, however, beware of these facts.

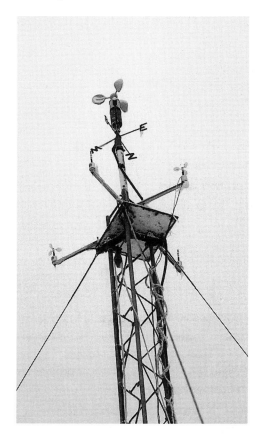

Figure 13.7. The cup anemometers and wind vane at the Brtish Antarctic Survey base at Faraday, on top of a 10 metre tower, are prone to icing and snow cover. However, since for much of the year the snow is dry and rime is not frequent, protection against them is not essential, although it is necessary to be aware of when sensor performance is impaired and to flag the data accordingly.

De-icing automatic stations

In Antarctica   Even in the case of fully automatic weather stations, operated remotely from bases in Antarctica, there is no widely used automatic ice and snow prevention or removal provision. However, in the remoter, colder regions where these stations are usually deployed, there is less of a problem since the lower temperatures reduce the frequency of severe icing or snowfalls. Rime can form, nevertheless, and snow does blow as spindrift, and so the problem still exists. But just as it is possible to infer from the data from manned stations when measurements are likely to be affected by ice and snow, so too is it possible with AWSs; the data give clues as to when to flag the measurements as suspect, for example when the wind is indicated as from a fixed direction for an extended period. But in marginal conditions, where some movement of the vane continues, errors may go undetected. Again, users of such data should beware.

What this illustrates is that one solution is to do nothing to prevent snow or ice forming on the sensors, but to be aware of when it is occurring and to quality-control and flag the data appropriately. But this does mean that the data will have gaps in them, and/or be inaccurate, and it is only feasible at locations where rime formation and wet or blowing snow do not occur for long periods. Having gaps in the data from these locations (as indeed from any location) means that long-term means and totals are not fully reliable.

By the use of heat   Techniques have been developed to combat the snow and ice problem automatically, the normal strategy being to use heat, mostly electrically generated. But large amounts of energy are necessary for this approach, owing to the size of the sensors, the high winds prevailing and the energy required to melt ice. Thus a cup anemometer (Hartley 1970) used 1.2 kW of heat, while a Pitot-tube wind sensor required 750 watts (Gerger 1972) to be kept ice-free. Gerger also describes a solarimeter heated by air blown over the dome. If each sensor is heated individually in this way, the power demands increase *pro rata*, and the large amount of power required is not usually available at remote sites, although it would be at Antarctic bases; it is surprising that more prevention has not been done where power is available.

Gerger (1972) reported on the technique of putting sensors in a housing that opens periodically to expose them for a short time, the housing being heated only at its opening point and internally (sufficient to keep the sensors just above freezing), thereby using less power. Alexeiev *et al.* (1974) noted the use of this technique for anemometers on ships, where adequate power is available, although our earlier comments about the problems of using anemometers on ships, even in non-icing conditions, must raise doubts about the usefulness of this.

A recent example of a heated enclosure is an AWS on Cairn Gorm in Scotland, developed by Heriot Watt University (Curran *et al.* 1977, Barton & Bothwick 1982, Barton & Roy 1983). In this AWS, sensors for measuring wind speed and direction, temperature and humidity are housed together in an insulated cylinder (Fig. 13.8), heated at its opening point and internally, and opened by an electric motor every 30 minutes for 3.5 minutes. This gives sufficient time to measure the mean wind speed and to detect peak gusts as well as allowing the temperature sensor time to attain equilibrium with the atmos-

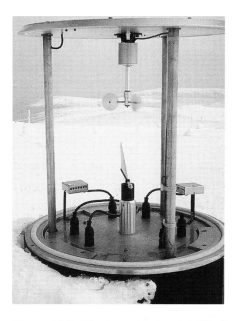

Figure 13.8. The 'opening box' AWS, designed by Heriot Watt University, is heated electrically (internally and at its opening point). Here it is seen open while one of its half-hourly readings is being taken; this exposes the anemometer (top plate), wind direction and temperature sensors (central base plate). The high altitude of the mountain (1245 m) makes it possible to telemeter its data to Edinburgh, 133 km distant, in one single UHF radio link.

phere before being read, yet it is a sufficiently brief period to avoid ice and snow accretion. Above freezing point, the heating is disabled. To keep all four sensors ice free requires an average power of only about 400 watts, a considerable reduction on earlier heated sensors exposed continuously and individually. But even 400 watts is beyond the resources of most remote sites. The Heriot Watt AWS is, however, quite large (1 m high by 0.75 m diameter) and power requirements could be reduced, perhaps considerably, if the design was miniaturised.

Indirect methods of de-icing   Non-heating techniques for ice prevention have been used in various applications, such as helicopter blades, aircraft wings, radar domes and ship's superstructures, but few have been applied to meteorological sensors. They have the advantage of requiring much less power to operate them than those based on heating. Jellinek (1957) and Jellinek *et al.* (1978) observed that ice adheres less well to hydrophobic materials, such as plastics, than to metals, but the usefulness of this for ice prevention is, in practice, minimal since although the strength of adhesion is certainly about ten

Figure 13.9. This pneumatically de-iced AWS was developed by the Institute of Hydrology and tested alongside the opening box station (Fig. 13.8), as well as in Antarctica for 12 months (on the right in Fig 13.6). It measures wind speed and direction with a drag anemometer (the top cylinder, see also Fig. 5.9), solar radiation (the sensor on top of the anemometer) and reflected solar radiation, temperature and humidity by means of sensors in the cavities on the underside of the plate.

times lower, it is still very large (about 2 kg cm^{-2}) (Sayward 1979). However, while ice adheres well to all rigid materials, irrespective of composition, in most cases it sheds easily from all that are flexed – perhaps just slightly more easily from hydrophobic materials. But if the ice has formed as numerous, unconnected small individual crystals, bending the surface may not remove them. These facts were used by the Institute of Hydrology in the design of a *cold regions* AWS (CRAWS; Strangeways 1985a, b).

A CRAWS consists of an aluminium plate 45 cm square, sandwiched between 1 cm thick foam rubber sheet, the whole being sealed in a polythene envelope (Fig. 13.9). Cavities in the foam-rubber house sensors measuring temperature (by PRT), humidity (by ion-exchange sensor), solar radiation and reflected solar radiation (by light-sensitive diode). Each cavity is protected by a material to suit its function, (reflective) mylar for temperature, (water-vapour-permeable) PTFE for humidity (as used also on buoys) and (diffusive) white polythene for radiation, all plastic and all flexible. On top of the plate, a drag-force anemometer (Chapter 5), encased in a flexible silicone-rubber cover, senses wind speed and direction. The whole structure is deployed on a pneumatic cylinder containing a piston and protected by rubber bellows. Air from an aqualung, controlled by a timer, periodically fires the piston, causing a sudden short vertical movement and thereby inducing a mechanical shock; the vented air is then directed so as to flex all the covers, the combined effect being to remove any ice that has formed or snow that has accumulated or adhered.

Nevertheless, both the Heriot Watt and Institute of Hydrology designs remain specialised and expensive. For wider application, it is possible to visualise further developments of these ideas, perhaps miniaturised and perhaps combining heating and pneumatics. These are open fields for further research.

References

Alexeiev, Jo. K., Dalrymple, P. C. & Gerger, M. (1974) Instruments and observing problems in cold climates. WMO Technical Note No. 135.

Banfield, E. (1976) *Antique Barometers*. Baros Books, Trowbridge, UK. ISBN 0 948382 04 X.

Barlow, D. (1994) From wind stars to weather forecasts: The last voyage of Admiral Robert FitzRoy. *Weather*, **49**, 123–8.

Barlow, D. (1997) The devil within: evolution of a tragedy. *Weather*, **52**, 337–41.

Barton, J. S. & Borthwick, A. S. (1982) August weather on a mountain summit. *Weather*, **37**, 228–40.

Barton, J. S. & Roy, M. G. (1983). A monument to mountain meteorology. *New Scientist*, 16 June 1983.

Begg, J. S. (1984) At Ben Nevis observatory. J. *Scottish Mountaineering Club*, **33**, 43–51.

Birch, K. G. & Pascal, R. W. (1987) A meteorological system for research applications – MultiMet. In *Proc. 5th Int. Conf. on Electronics for Ocean Technology, Edinburgh*, pp. 7–12. Institute of Electrical and Radio Engineers (IERE) Publ. 72.

Chenoweth, M. (1996) Nineteenth-century marine temperature data: comments on observing practices and potential biases in marine datasets. *Weather*, **51**, 280–5.

Clarke, C. S. & Painting, D. J. (1983) A humidity sensor for automatic weather stations. In *AMS Proc. 5th Symp. on Meteorological Observation and Instrumentation, Toronto*.

Curran, J. C., Peckham, G. E., Smith, S. D., Thom, A. S., McCulloch, J. S. G. & Strangeways, I. C. (1977) Cairn Gorm summit automatic weather station. *Weather*, **32**, 60–3.

EGOS (1996) European Group on ocean stations. EGOS brochure, CMR 05/1996, WMO/IOC.

Eumetsat (1995) Data collection system, user guide. EUM UG 02, pp. 29–34.

Folland, C. K. (1988) Numerical models of the raingauge exposure problem, field experiments and an improved collector design. *Quart. J. Roy. Met. Soc.*, **114**, 1485–516.

Folland, C. K. & Parker, D. E. (1995) Correction of instrumental biases in historical sea surface temperature data. *Quart. J. Roy.* Met. Soc., **121**, 319–67.

Gerger, H. (1972) Report on the methods used to minimise, prevent and remove ice accretion on meteorological surface instruments. (WMO) CIMO-VI/Doc. 35, Appendix A.

Hartley, G. E. W. (1970) A heated anemometer. *Met. Mag.*, **99**, 270–4.

Hasse, L., Grossklaus, M., Uhlig, K. & Timm, P. (1997) A ship rain gauge for use in high wind speeds. *J. Atmos. and Ocean. Tech.*, **15**, 380–6.

Jellinek, H. H. G. (1957) Adhesive properties of ice. US Army Snow, Ice and Permafrost Research Establishment, Corps of Engineers, Research Report No. 36.

Jellinek, H. H. G., Kachi, H., Kittaka, S., Loe, M. & Yokota, R. (1978) Ice-releasing block-copolymer coatings. *Colloid Polym. Sci.*, **256**, 544–51.

Kenworthy, J. M. & Walker, J. M. (1997) Colonial observatories and observations: meteorology and geophysics. In *Proc. Conf. Roy. Met. Soc., University of Durham*, April 1994. ISBN 0307 0913.

McConnell, D. (1988a) The Ben Nevis observatory log-books, part 1. *Weather*, **43**, 356–62.

McConnell, D. (1988b) The Ben Nevis observatory log-books, part 2. *Weather*, **43**, 396–401.

Meteorological Office (1995) Marine observers handbook. HMSO, London.

Meteorological Office (1997) UK Meteorological Office moored data buoy: technical description. Meteorological Office Technical Description K-27-TD, Issue 1, August 1997.

Parker, D. E., Folland, C. K. & Jackson, M. (1995) Marine surface temperature: observed variations and data requirements. *Climatic Change*, **31**, 559–600.

Paton, J. (1954) Ben Nevis observatory. *Weather*, **9**, 291–308.

Roy, M. G. (1983) The Ben Nevis meteorological observatory 1883–1904, Part 1. Historical background, methods of observation and published data. *Met. Mag.*, **112**, 218–329.

Sayward, J. M. (1979) Seeking low ice adhesion. Cold Regions Research and Engineering Laboratory, Special Report: No. 79–11.

Schutze (1883) The arctic meteorological station on the Lena (River). *Nature*, **28** (59).

Strangeways, I. C. (1985a) The development of an automatic weather station for use in arctic conditions. PhD thesis, Reading University, Department of Meteorology.

Strangeways, I. C. (1985b) A cold regions automatic weather station. *J. Hydrol.*, **79**, 323–32.

Strangeways, I. C. (1996) Back to basics: The 'met. enclosure': Part 2(b) – Raingauges, their errors. *Weather*, **51**, 298–303.

Warner, T. B. (1972) Ion-selective electrodes. Properties and uses in sea water. *J. Mar. Technol. Soc.*, **6**(2), 24.

Yelland, M. J., Taylor, P. K., Consterdine, I. E. & Smith, M. H. (1994) The use of the inertial dissipation technique for shipboard wind stress determination. *J. Atmos. Tech. (Am. Met. Soc.)*, **11**, 1093–108.

14

Remote sensing

This is but a brief review of a complex subject and is intended to give an overall impression rather than a detailed account. The aim is to clarify what can be sensed remotely, and to put remote sensing into context with ground-based measurements.

What is remote sensing?

Remote sensing (RS) is *not* the transmission of data from *in situ*, ground-based sensors at a remote site to a distant base. That is telemetry (Chapter 12).

Whereas *in situ* measurements are made by sensors in direct contact with the variables they measure, RS measurements are made entirely by sensing the electromagnetic radiation reflected from, or emitted by, the surface of the earth and its atmosphere. Astronomy provides a good example of RS (apart from the spacecraft that have soft-landed on other planets), and photography and, more recently, electronic imaging have become the sensors and recorders of astronomy.

A *platform* from which RS measurements are made can be simply a mast on the ground, an aircraft, balloon, rocket or spacecraft in orbit around the earth. It is the last that has had the biggest impact on RS and so satellite RS is the main concern of this chapter. However, many of the instruments and techniques used on spacecraft are also used on the other platforms, particularly aircraft. But aircraft are expensive to fly and so are used mainly where an occasional or single measurement is needed and where a local, detailed, low altitude view is necessary. Many satellite images are free, and regular repeat-images are available covering the whole planet.

An undoubted attraction of RS is its low cost. Satellite images are available free of charge and in real time for the price of a receiver (Chapter 12). While a primary data user station (PDUS; see Chapter 12) is not as cheap as a Wefax receiver, compared with the price of installing and maintaining *in situ* instruments over large areas it is very cheap indeed. The large price differential tempts cut-backs in ground-based observations, but this would be unfortunate, and ill-advised, for RS cannot replace ground observations, indeed it relies on them – as will be shown.

The electromagnetic spectrum

Figure 14.1 illustrates the electromagnetic spectrum, with the extreme ends (low radio frequencies and gamma rays) omitted to save space and also because these extremes of frequency are not used in RS from satellites as yet. The figure brings together information regarding the wavelength, frequency and photon energy, the names of the radiation, the spectrum of the sun and the earth's long-wave infrared spectrum (see also Fig. 2.2), along with an indication of the types of instrument used to measure the various bands. An approximate indication of the bands of radiation that are absorbed and transmitted by the earth's atmosphere are also shown. These 'windows' are important, since RS of the surface is only possible from space in the bands that are not absorbed by the atmosphere, in particular the visible and the two infrared windows A and B, where transmittance is high, shown in Fig. 14.1.

The division of radiation into ultraviolet, infrared, microwave, etc., does not, of course, indicate a sudden change in the nature of the radiation at the point where the name changes. Except in the case of visible light the division is partly for convenience, and various break-points will be found in different writings; the ones used here are typical and as good as any other. The division of radar into K, X, C, S, L and P bands, as shown in the figure, is also just one of several possible frequency divisions.

Measuring the radiation

Most RS sensors measure the natural radiation from the sun reflected from the earth and its atmosphere and also the long-wave infrared and microwaves originating directly from the earth and the atmosphere by virtue of their constituting a *black body* at a temperature above absolute zero. Sensor systems measuring radiation from these sources are referred to as *passive*. Figure 14.2 illustrates the various routes taken by, and sources of, the radiation. It is also possible for an instrument to have its own source of radiation, in particular microwaves, generated by a radar transmitter on board the platform, the sensor detecting the reflected radiation. These methods are called *active* (Blyth 1981).

Film cameras

For RS in daylight from manned platforms (including spacecraft) in the visible and near-infrared bands, direct photography on film offers many advantages by providing high-image resolution (10 million bits on a 70 by 70 mm film) with modern optics and photographic emulsions, at relatively low cost.

The spectrum from 0.3 to 0.9 μm can be recorded by film, the shorter-wavelength threshold being limited by the transmittance of glass lenses and the longer-wavelength threshold by the sensitivity of the film emulsion. Quartz lenses can extend the spectral range, and mirrors can be used in place of lenses to avoid the losses that occur through glass.

Filters can be used to separate the image into several discrete bands; typically four cameras are operated together, fixed to a frame to ensure good image alignment, each sensitive to a specific band. The usefulness of splitting images into several wavebands is that it is often possible to identify materials (water, vegetation and soil for example), or their condition, by their different spectral reflectance. Film cameras are widely used from aircraft and from manned spacecraft.

Vidicon tubes

Vidicon cameras are well adapted to automatic, remote use, having the capability to produce electrical signals that can be digitised, stored and telemetered back to ground. However, while these were used widely in the 1970s and 80s, they have been largely superseded by scanning radiometers and so will not be described here.

Non-scanning radiometers

Visible wavelengths   Visible-light radiometers are fairly simple and measure the radiation within a fixed angle of view, either over a wide angle, averaging the whole scene (as in camera exposure meters) or in a highly focused way, using lenses for spot measurements. The light is directed onto a sensor such as a photoresistive (photoconductive) cell of cadmium sulphide (sensitivity 0.5–0.9 μm) or a photodiode of the type used to measure solar radiation (Chapter 2). Semiconductor radiation sensors respond to individual incoming photons and are thus known as quantum sensors, responding rapidly to variations in radiation level – although they do not have a response to very high frequencies because of the relatively slow decay of the charge carriers.

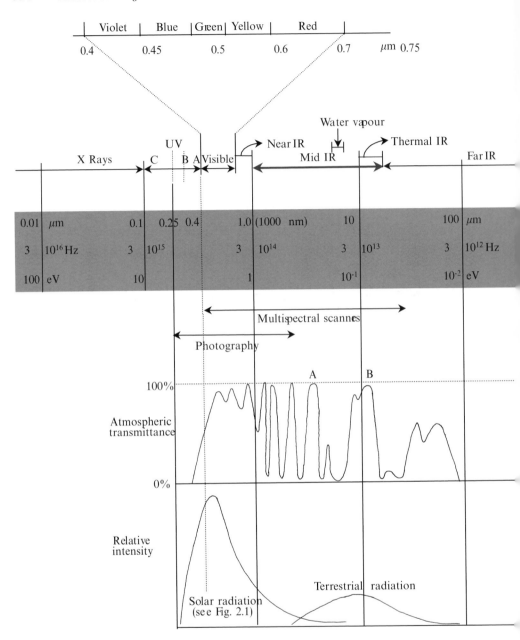

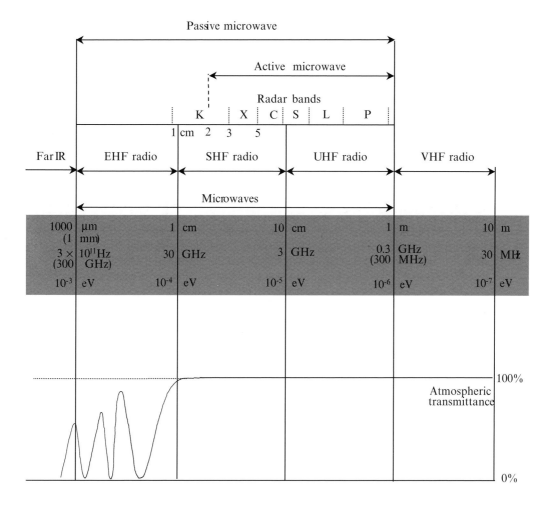

Figure 14.1. The electromagnetic (em) spectrum, from X rays to HF radio, showing the subdivisions by name, frequency and wavelength, with approximate curves of atmospheric transmittance and the radiation spectra of the sun and the earth. $\lambda = c/f$, where λ is wavelength in m, f is frequency in Hz and c is the speed of light in a vacuum, 3×10^8 m s^{-1}. In each part of the em spectrum, the photon energy is given in electronvolts, eV; 1 eV is equal to 1.6×10^{-19} J.

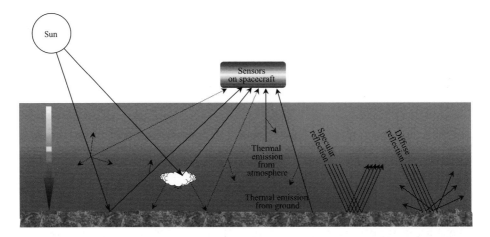

Figure 14.2. Because of the presence of the atmosphere, solar radiation may have followed a complex path of multiple reflections, scattering, diffusion and absorption before arriving at a sensor, such as a radiometer on a satellite. The solid lines show the direct and reflected solar radiation; the dotted lines show diffuse radiation from a cloud and radiation scattered by the atmosphere. Natural thermal emission originating from the warmed ground and atmosphere is also indicated. The general absorption of solar radiation as it passes through the atmosphere is shown by the large arrow on the left. How radiation is reflected at the ground depends on the nature of the surface and can be *specular* when the surface is smooth or *diffuse* when it is rough, the processes varying with wavelength and angle of incidence. At water surfaces, ultraviolet is reflected and near-infrared absorbed, while the blue-green spectrum achieves maximum penetration, although water waves complicate the situation. All these factors make remote sensing a difficult and complex task.

Infrared wavelengths   There is a range of pure and doped semiconductor photoresistors covering the spectrum from 0.5 to 20 μm. For example, in place of cadmium sulphide a sensor of mercury–cadmium telluride, kept at a controlled temperature, can be used in an infrared radiometer. In some radiometers, a rotating shutter passes the incoming radiation directly onto the sensor, alternating it with radiation reflected from an internal calibration reference cavity at a controlled temperature. The difference in reading between the two sources is proportional to the temperature of the incoming radiation; the internal reference also counteracts any drift in sensor and electronic characteristics.

Microwaves   Although visible and shorter infrared radiation can be collected by optical methods, in the microwave bands an antenna is used, focusing the radiation onto a horn which then passes it to a diode detector. Very low levels of natural radiation emanate from the ground in the microwave region and, by using a parabolic dish, it is possible to boost the signal by increasing the collecting area. The size of the dish also influences the spatial resolution obtained, this being dependent on the distance from the target and the beam width. The beam width is proportional to the wavelength and inversely pro-portional to the diameter of the dish. In practical terms this means that if the radiometer is on board a satellite at a height of 500 km and the dish is 1 m in diameter, then at a wavelength of 3 cm the resolution will be a spot 18 km in diameter (Blyth 1981). While this might be useful for obtaining average measurements over large areas, it is clearly no good for finer observations. (If flying in an aircraft at 5 km altitude, the resolution of the same dish is about 180 m.) Radars with one dish working in this way are known as *real-aperture* systems, to differentiate them from *synthetic-aperture* radars, which have better resolutions; see below.

A microwave radiometer measures what is called the *brightness temperature* of the ground, which is a function involving both the temperature of the ground and its emissivity, the latter being largely dependent on the ground's dielectric constant (see Chapter 9, the subsection on the capacitance probe); surface texture also affects the emissivity (Fig. 14.2). There is, therefore, no direct and simple equivalence between the measured radiation and ground temperature.

Spectroscopy

By inserting a light-dispersing element, such as a prism, into the radiation path of a visible or infrared radiometer, the beam can be split into discrete spectral bands and the intensity of each measured. By using a variable interference filter in place of a prism, it is possible to divide up a wide spectrum into very narrow (5 μm wide) bands, and if a reference source of radiation is also included in the instrument, precise calibration of the intensity of each band is possible.

But radiometers only look at one spot at a time, or encompass very large views, and so are limited in what they can achieve. To do more, some form of scanning of the scene is necessary to produce images.

Scanning radiometers

To combine the potentially high resolution of a visible or IR radiometer with wide areal coverage, it is necessary for the radiometer to scan the view, both vertically and horizontally.

Scanning radiometers in polar orbits   In a geostationary satellite such as
Meteosat, both the X- and Y-axes have to be scanned, but if the satellite is
moving relative to the ground, such as one in a polar orbit, the forward
(along-track) motion of the satellite provides the X-scan, only the Y-axis
having to be scanned by the instrument. If the along-track motion of the
satellite is used, the frequency of the scan and the width of the scan strip have to
be chosen so as to take account of the altitude and speed of the platform, to
ensure that adjacent individual strips neither overlap each other nor have gaps
between them. In an aircraft, there can be a problem with distortion, owing to
the increasing angle of incidence as the angle of the scan increases to left and
right, although correction can be made for this. From space, distortion is less,
owing to the greater height (Wong 1975).

The most usual way of scanning the ground is to rotate a mirror at 45° to the
horizontal at several hundred rpm by means of a motor, the radiation being
reflected into the system's optics, typically a small Cassegrain telescope, as
shown in Fig. 14.3. When the mirror is pointing downwards it scans the earth;
as it looks up it reflects radiation from an on-board calibration source, in a
similar way to that in a fixed radiometer. It can also look at the low temperature of deep space as an additional reference at a few degrees above absolute
zero. In some designs the mirror vibrates rather than rotates, but rotation is
more usual.

The same types of photoconductive sensor are used in scanning radiometers
as are used in fixed radiometers: cadmium sulphide for the visible and for part
of the near-IR, and mercury–cadmium telluride for the near- and mid-IR. The
most usual IR bands to be measured are those at 3–5 µm and 8–14 µm, since
there are 'windows' at these frequencies through the atmosphere (marked A
and B in the transmittance graph in Fig. 14.1).

The need to use a large dish in the microwave region to get sufficient
resolution was alluded to above, and this problem is exaggerated further if the
dish has to move, as in a scanning system. From an aircraft the problems are less,
in part because the resolution is better due to the low altitude, and in part
because planes are larger than RS satellites. On aircraft it is also possible to use
three dishes rotating on a common axis, switching so as to receive from just the
forward-looking dish – it is usual to look slightly forward, rather than straight
down, in the case of microwaves as this helps to identify the nature of the surface
better, through polarisation effects. It is also possible to use two dishes spaced,
say, n metres apart and rotated in synchronism, which, with suitable computer
processing, can make the combination look like a dish n metres in diameter,
thereby greatly improving resolution. However, on small spacecraft this arrangement is too large to be practical and special aerials known as *phased array*

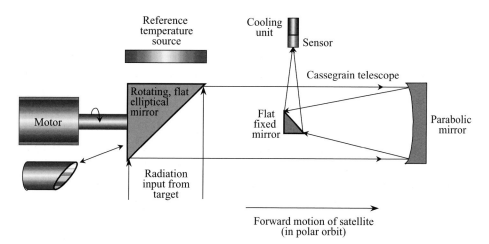

Figure 14.3. A scanning radiometer aboard a satellite in a polar orbit uses the forward motion of the satellite to generate the *X*-scan, a rotating mirror generating the *Y*-scan. The collected light is directed to a small Cassegrain telescope, which focuses the radiation onto the sensor. When facing upwards, the rotating mirror receives radiation from a body of known temperature for calibrating the system. This body can be deep space.

antennae, which can reproduce the properties of large antennae, have been developed, allowing higher resolution. They have no moving parts, and this allows fast scanning and alterations to the shape and width of the beam as well as control of the polarisation sensitivity, all these characteristics being controlled electronically. But they are expensive and (it is understood) none are yet in operation on spacecraft.

Scanning radiometers in geostationary orbits   Aboard a geostationary satellite such as Meteosat, where there is no motion of the satellite relative to the ground, radiometers have to scan in both the *X*- and *Y*- directions. To achieve this the whole satellite rotates on its axis at 100 rpm, the spin creating the *X*-scan, while a mirror steps down by a very small angle at each rotation, giving the *Y*-scan. In this way, over 25 minutes, the satellite scans all the visible globe, producing 2500 scan lines. (To compensate for rotation of the whole satellite, its antennae have to be electronically switched, continually – so that radio communication is maintained with the earth even though the satellite is rotating.)

Multispectral scanners

Multispectral scanners (MSSs) represent the next step-up in complexity, combining a scanning radiometer with spectroscopy. Instead of radiation from the

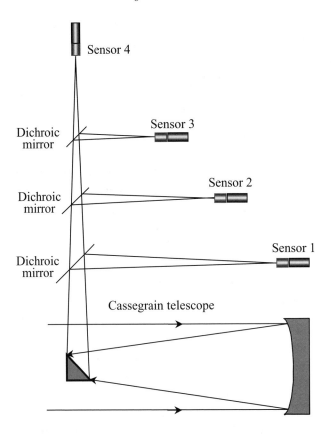

Figure 14.4. In principle, a multispectral scanner is a scanning radiometer in which the incoming radiation is split into several separate beams by dichroic mirrors, with a sensor associated with each that responds to a different wavelength, perhaps with its frequency discrimination further enhanced by filters.

Cassegrain telescope being directed to a single radiation sensor, it is passed through several dichroic mirrors. The radiation from each is focused onto a separate sensor, sensitive to a particular part of the spectrum, possibly with an additional filter to modify the sensor's range of spectral response (Fig. 14.4). *Thematic mappers* (TMs) are similar to MSSs but measure narrower bands of radiation, generally seven bands compared with the scanner's four, to higher resolution (20 compared with 80 metres).

Active microwave sensing

All the above remote sensing (RS) instruments measure radiation that is of natural origin. It can be very useful, though, to actively emit radiation from the platform and measure the reflected signal. This is most commonly done in the microwave bands using a radar transmitter. The frequency used is usually fixed on any one platform, but can be anywhere in the active microwave range, as shown in Fig. 14.1. There are advantages, however, in using multiple radar frequencies – for the same reasons that it is useful to employ a number of visible bands.

In the most commonly used radar technique, a radar pulse is emitted sideways and downwards from the platform, at right angles to the direction of travel, at a pulse repetition rate of from 1000 to 10000 pulses per second. A fan-shaped volume of radiation is generated as the pulse travels outwards from the antenna (Fig. 14.5), the pulse arriving at the ground at increasingly greater distances and at increasingly later times, as it propagates outwards at the speed of light. At the moment shown, the region of ground returning the signal is the shaded area; the receiver on the platform thus detects returned radiation from each such area sequentially, using the same antenna as that which transmitted the pulse, switched to the receiving mode. The amplitude, phase, polarity and scatter of the returned signals all give information on the topography, texture and nature of the surface. The *angular spread* of the transmitter, β, and the *depression angle* θ determine the area of ground illuminated. The width of the ground g across the swath area, depends on the width of the pulse, τ, and on θ, which is usually set at between 30° and 60° to obtain optimum resolution.

But the resolution of the ground area covered in the direction of flight, W, is limited by the antenna beam width, which is a function of antenna size and the frequency used. In most real-aperture systems (see earlier in the chapter, in the subsection on scanning radiometers), the across-track resolution g is ten times better than W. As already discussed (under passive microwave radiometers) it is possible to simulate a larger antenna by using two dishes spaced some distance apart. In the case of active radar instruments, this 'second' antenna is actually the same antenna – but, having moved further down the flight-path, it takes overlapping measurements (Fig. 14.5) separated by time. In this way, the same effect is produced as having two separate antennae spaced by the distance D. This is referred to as a *synthetic aperture radar* (SAR). How the information is merged to produce this artificially narrower beam is mathematically complex, but the end product is a resolution in the order of 25 to 50 metres from a polar orbit.

The most-used wavebands

The largest concentration of measurements from satellites is in the visible region (0.4 to 0.75 μm) with only slightly fewer being made in the near infrared

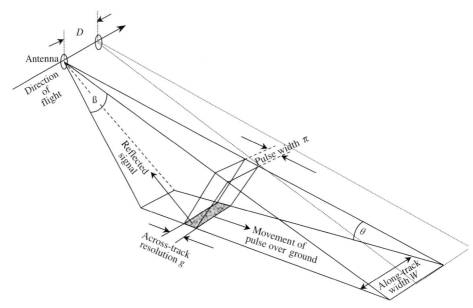

Figure 14.5. In a common form of active remote sensing, a radar pulse is emitted sideways and downwards, the reflected signal being returned from the shaded area at increasingly distant points as the pulse travels outwards. To make the along-track resolution width, W, as good as that across the track, g, a technique known as *synthetic aperture radar* is used. By timing pulse emissions (relative to the speed of travel), the scanned areas can be made to partly overlap. Then, by receiving echoes from the same area but originating from two positions, the effective diameter of the radar antenna can be artificially increased (through complex processing) to D, thereby effectively narrowing the beam width and so increasing the resolution.

(0.75 to 1.1 μm). The next most common band is in the thermal infrared window marked B in the transmittance graph of Fig. 14.1 (10.5 to 12.5 μm). Fewer sensors operate in the window marked A (3.5 to 4.0 μm). In the region between 6 and 7 μm water vapour radiates strongly, and a few satellites (such as Meteosat) make measurements of humidity by sensing this frequency band. There is no atmospheric window in this band, since water vapour not only radiates at these frequencies but inevitably also strongly absorbs. Meteosat (which is typical of its type) measures in the three bands 0.45 to 1.0 μm (visible and near IR), 5.7 to 7.1 μm (water vapour IR) and 10.5 to 12.5 μm (thermal IR).

In the microwave region, there is a concentration of satellites operating in the C-band using SAR active sensing techniques, with another group in the X- and K-bands acting as passive scanning radiometers. But there is continuous development and change with the launch of each new satellite.

Image reception and dissemination

For retrospective use, images may be obtained from the satellite operators, either as printed images or as digital data. For use in real time, however, the images must be received directly from the satellites. Something has already been said in Chapter 12 about the latter, with regard to Meteosat, and how image reception is associated with the reception of data telemetered from DCPs. PDUSs and SDUSs (*primary and secondary data user stations*) receive image data directly from Meteosat, the former as high-quality digital data, the latter as lower-quality analogue Wefax data (known as APT, *automatic picture transmission* – from satellites in polar orbits). Figure 12.16 shows a small dish for the reception of both DCP data and Wefax images from Meteosat. Meteosat images are divided into nine square sections showing different parts of the earth's disc as well as the whole disc. In addition there are more frequent repeat-images of North Africa and Europe. The other geostationary satellites produce similar images of their area (Chapter 12).

The dissemination of satellite images is explained fully in several Eumetsat publications (see the reference list for Chapter 12), and how images can be received by the amateur directly from the various satellites is well documented by Harris (1996). Further details need not be followed here.

Limitations of remote sensing

Spatial and temporal restraints

The orbits of satellites have already been dealt with in Chapter 12. It is useful, however, to condense relevant facts into a brief restatement regarding image resolution (spatial and temporal). There are two types of orbit in general use for environmental, weather (and communication) satellites, namely geostationary and polar.

In the former, satellites are operated in orbits at a distance of approximately 35 900 km from earth over the equator, giving them an orbital period of 24 hours. They circle the earth in the same direction as the earth rotates, thus with the same view of earth all the time. This, and the 25 minute scan-time of the spin-radiometer (see above), allows repeat images (of the whole disc) every half hour. But the spatial resolution is 2.5×2.5 km square, and even this is only achievable at the sub-satellite point (nadir), at which the ground viewed is directly beneath the satellite. As the field of view extends away from the nadir point, the pixels become increasingly larger, elongated and distorted, so that at the limits of view the ground is seen at a grazing angle. The higher polar regions cannot be seen at all.

Polar orbits may be true polar (at 90° to the equator) or skewed at various oblique angles. Satellites in polar orbits are low-flying (a few hundred to a thousand kilometres) and so, being closer to the ground, they collect images with a much higher spatial resolution than those from geostationary satellites. In the case of the Spot satellite, for example, the pixels are as small as 10 metres square, while for Landsat the resolution is from 30 to 120 metres square, depending on the instrument used, multispectral scanner (MSS) or thematic mapper (TM). Because the ground is so close, earth-curvature is small, and so the pixels are more or less the same size and shape across the full view. But while polar orbits give much better spatial resolution than geostationary orbits (if the swath width is small), the price paid is a reduced time coverage. In the case of satellites in true polar orbits, the poles are in view at each pass (14 times a day in the case of Argos). As the satellite moves from the poles towards the equator, however, overpasses of any given section of ground naturally become increasingly less frequent, and at the equator (although there are still 14 passes a day somewhere over the equator) there may be many hours without an image of any particular area. With some satellites, repeat-images may be days apart, even up to two or more weeks.

Clouds and night

If clouds are the subject of interest, then their presence is to be welcomed, but if it is some aspect of the surface beneath the clouds that is of concern, clouds are a problem in the visible and the infrared spectrum. In mid-latitudes and in the tropics, clouds can obscure the view frequently and for long periods. If the measurements are also made by an infrequently passing polar satellite, opportunities to observe the ground might be weeks or months apart. In the visible and non-thermal IR spectrum, night time also restricts observations to half the day. With rapidly changing variables, such as many of those of concern to meteorology and hydrology, both these limitations are obviously a major problem. For example, it would be impossible to reproduce the performance of a conventional meteorological site using remote sensing if measurements were only available every few days. There are, of course, some things that do not need to be measured frequently because they change relatively slowly, such as ice- and snow-cover and land-use. But if high-resolution observations are required at frequent intervals in the visible and IR spectrum then cloud, night time and infrequent overpasses would make them impossible. These factors alone limit greatly the usefulness of RS in many applications.

What can be measured remotely?

It is not the aim of this chapter to enquire how every variable is measured remotely; that has been the subject of many publications (for example, Barrett & Curtis 1982, Barrett & Herschy 1986, Blyth 1981, Farnsworth *et al.* 1984, Rango *et al.* 1983, Schultz & Barrett 1989 and Stewart *et al.* 1998a, b). It is a highly complex subject, with different methods for each variable and often several different methods for one variable. The impression often gained is that RS can do much more than it actually can, probably due to its undoubted success in the simpler qualitative applications. Quantitative applications, however, are quite a different matter.

Qualitative observations

Clouds, snow, vegetation and geology   The most obvious information from satellites is that regarding cloud-, snow- and ice-cover, or semiquantitative data such as geological formations, crop development, vegetation maps and conditions indicating pest risk. The majority of RS applications are of this type. Indeed images of clouds are perhaps one of the greatest benefits of satellite RS, giving information never before available. Eumetsat state that the main mission of Meteosat is to generate 48 cloud images a day for use in short-period forecasting, the visible channel producing images in daytime and the infrared channels during both day and night. Sequences from all three channels are used to determine the motion of clouds, from which the apparent wind speed and direction can be deduced. In addition, the *International Satellite Cloud Climatology Programme*, part of a WMO research project, produces global maps of monthly averages of cloud amounts. The same images can also track weather fronts and developing weather systems.

Measuring open water   Open water such as lakes can also be easily detected, using the near-IR bands which are strongly absorbed by water but highly reflected by plants. This difference can be detected in a yes/no way, allowing images to be analysed automatically. It is not so simple, however, in the case of rivers because their width is much less than that of a lake, and will normally be less than one pixel wide, even from a polar orbit. Here, indirect evidence is often sought instead, such as the tell-tale signs of changes in vegetation in the proximity of water, but these can only be identified by eye, not automatically. Forest fires can also be usefully detected. Altitude and slope of the land are fairly straightforward to measure using active microwaves (synthetic aperture radar, SAR). Significant wave height (see p. 315) can also be well estimated by radar (Carter 1999).

None of these variables, however, needs to be expressed as a precise quantity, apart from altitude, which is simple to measure with radar, and so the images are (relatively) easily interpreted. But there are many variables that are much less simple to measure remotely because precise values *are* required and because of the spatial and temporal restraints posed by satellite orbits, clouds and night.

Quantitative measurements

Algorithms   Although the geographical extent of a lake may be defined precisely, as described above, as soon as an attempt is made to measure, for instance, the lake's depth, its temperature or the quality of its water by RS, difficulties start. Many such variables are far less amenable to remote sensing than is a simple yes/no situation, or a situation where only semiqualitative information is involved, because the electromagnetic spectrum reflected from, or emitted by, the earth's surface and measured by the satellite instruments is rarely directly related to the variable of interest. To overcome this difficulty, intermediate models (algorithms) have to be developed to interpret the remote measurements, and these algorithms can be quite complex and may not be very precise, or they may only be applicable to a particular location, or to a particular set of circumstances, or to one season. The largest effort being put into RS by individual research groups is concerned with developing these algorithms. (Spacecraft sensors are also being improved, as are the general products from, for example, Eumetsat, but these are the concern of the satellite operators and largely beyond the user's control.)

The topic of algorithms is extensive and complex and cannot be elaborated on here, although it is useful to list just a few of the commoner variables that are measured remotely, and to note very briefly the salient points regarding the indirect methods used. This should throw light on the general nature of the problems encountered and on their solutions, and also on why the accuracy of such measurements can still be quite limited. It also demonstrates the importance of ground-truth measurements to RS, as we now discuss.

The importance of ground-truth measurements   It is possible to ameliorate both the lack of any direct correlation between the measured radiation and the variable itself, as well as the effects of atmospheric interference on the passage of the radiation from ground to satellite, by making *in situ* 'ground-truth' measurements. (Atmospheric interference is one of the problems in satellite RS, being less apparent at the lower altitudes flown by aircraft; it is the same problem, but reversed, that astronomers encounter in working from the ground.) Ground-truth measurements, combined with the RS quantitative data via the algorithm, greatly increase the accuracy of the RS observations. Indeed it is this combination of ground-truth and remotely sensed data that gives the best results in RS – sometimes the only useful results. Ground-truth combined with RS also allows the precise *in situ* measurements from the ground-based instruments to be extrapolated to give better estimates of the variation between point observations and thus a better areal average. The two techniques thus assist each other in both directions. The combination of *in situ* with remote data is widely used, being applied to most of the variables that hydrologists and meteorologists, for example, need to measure. Ground-truth is also essential for the verification of algorithms. That NASA have gone to the great lengths of soft-landing instruments on other planets to make *in situ* measurements bears out the importance of ground-truth.

There is an additional problem, referred to as *scaling up*, which concerns the compatibility of a single point measurement made on the ground with the often much larger areal average measurement produced by many RS instruments, perhaps over a 25 square kilometre area; and there is the problem of being certain of the precise geographical location of the RS pixel relative to the ground observation. In addition there are the questions of temporal scaling-up from instantaneous satellite measurements to daily values and of ensuring that the RS measurements apply to the same instant that the ground-truth measurement was made. Furthermore, there is the question of the scaling-up of models (Stewart *et al.* 1998) – can models developed for use over a few square kilometres be expected to be correct on the normally much larger scale of RS?

Snow and ice The presence and extent of snow and ice is one of the few hydrological variables that can be measured from satellites on an operational basis, because, like open water, snow and ice can be recognised on a yes/no basis. The main problem is differentiating snow from cloud, but several methods have now been developed, such as that of Ebert (1987), which uses an algorithm to differentiate between the two based on albedo thresholds; Dozier (1987) describes a method using the 1.55 to 1.75 μm channel of Landsat, in which snow exhibits a much lower reflectance than cloud. But manual methods, looking at changes in clouds and the ability to identify ground features such as roads and forests, are also used. However, it is much less practicable to measure the depth of the snow or its equivalent water content than to measure its presence and extent.

Radiation   Solar radiation arriving at the ground evaporates water and heats the atmosphere and ground. It is, therefore, an important quantity. It is also one of the few variables that can be determined quantitatively and with reasonable accuracy by RS. A satellite radiometer measures the *short-wave* (solar) radiation being reflected back to space from the atmosphere and from the ground. Subtracting this measurement from the (precisely known) amount of incoming radiation just outside the atmosphere gives the amount reaching the earth's surface. This value is usually expressed as the fraction of radiation that would have been received under clear-sky conditions, this being a function of sun elevation and atmospheric turbidity. To take these into account as well as the multiple reflections, absorption and scattering that occur throughout the depth of the atmosphere, Möser & Raschke (1983) derived a polynomial equation giving the incoming radiation under clear-sky conditions for varying values of sun elevation, which incorporated coefficients based on atmospheric visibility data. Using this method, tests from a project in Mexico (Stewart *et al.* 1999) demonstrated that the error, averaged over a six-month period, was around 9% for daily values. If just clear days were included the error fell to 5%. This is a very good result for RS.

Temperature   As an example of the lack of direct equivalence between the radiation measurement and the variable, consider the same lake whose areal limits were simple to define (see above in the subsection on qualitative observations). Radiation measured in the thermal infrared gives an indication of the temperature of the lake, but only of the water surface, with no direct way of measuring the temperature even just a few millimetres below the surface. The situation is further complicated by the state of the surface, whether it is placid or wave-covered. The same is true of soil temperature; it is the temperature of the very top surface of the soil that is sensed. There is no direct or simple way of knowing what the temperature of the air a metre above the ground is (as is determined by an *in situ* thermometer in a Stevenson screen, for example) because the ground temperature is rarely the same as that of the air above it and in fact can differ considerably at times. In the same way, the temperature just beneath the surface of the ground is also unknowable directly from RS measurements. These other temperatures have to be inferred indirectly through a knowledge or estimate of other variables, processes or conditions, and the result is inevitably not as accurate as a primary measurement of the temperature of the surfaces. The surface-temperature measurements themselves are also likely to be in error owing to the passage of the radiation through the atmosphere – even when the measurement is made through one of the two main mid-infrared atmospheric windows (A and B, Fig. 14.1). Methods have been developed to correct for atmospheric effects, but they can never be completely successful. The same problem exists in making sea-surface-temperature measurements. Just the skin temperature is sensed and, despite the name, this is not exactly what is required (Harris *et al.* 1998). The *in situ* sea-surface temperature, measured by a ship or a buoy, is that of the water about one metre below the surface (Chapter 13).

If a profile of the temperature through the depth of the atmosphere is required then a different technique can be used in which a microwave radiometer measures the temperature-dependent thermal emissions of oxygen. The atmospheric concentration of oxygen is relatively constant in space and time, and since it both absorbs and emits at frequencies of around 60 GHz, it provides a stable temperature tracer. Measurements made at 53.74 GHz give an indication of temperature in the middle troposphere and those at 67.95 GHz, the lower stratosphere. These can be combined to give an indication of the average temperature of the lower atmosphere, weighted mostly in the region of 1.5 to 5 km altitude. These measurements have been used in the debate about global warming (Jones & Wigley 1990, Spencer & Christy 1990).

Rainfall   The techniques for sensing rainfall remotely fall into three categories: measurement of the rain as it falls, estimation of how much rain may be falling and observation of the resultant ground condition after rain has fallen.

Weather radar, which is earth-based RS, has already been described (Chapter 8). It is complex and expensive (even on the ground) and not practicable from a satellite as yet. Also, since one occasional spot observation is not very useful in the case of rainfall measurement, any such direct measurement of rainfall would have to be from a satellite in geostationary orbit. This is not currently technically feasible. So it is necessary to look at how precipitation can be inferred indirectly.

Many indirect methods of estimating how much rain is likely to be falling have been developed over the last 20 years; these involve looking at the clouds themselves from above, rather than at the rain falling from beneath them. These techniques generally use the satellite measurements of cloud-top brightness and/or temperature and/or texture, or the rate of cloud growth, with use being made of the visible, infrared or both bands, coupled with other 'predictors' such as ground-truth raingauge surface measurements.

There are many such techniques, a well-known one being the operational method developed by Reading University in Niger using Meteosat images, which estimates ten-day and thirty-day rainfall totals (Dugdale & Milford 1986). The method relates the duration of cold cloud-tops, as measured by the satellite's thermal infrared channel, to the precipitation. It is applicable only to

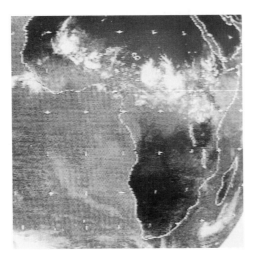

Figure 14.6. Cloud-top temperatures can be used to assess the possibility of and probable intensity of rainfall below convective clouds of the type that occur in many equatorial regions, as seen here in a Wefax image of central Africa from Meteosat.

convective cloud systems and is most useful for the production of longer-term rainfall summaries. However, a large part of the rainfall over Central Africa originates from such deep convective cloud. Figure 14.6 shows a section of a whole-disc Meteosat thermal infrared image (10.5 to 12.5 μm), in which such cloud formations can be seen over Africa. However, the more detailed images produced by Meteosat are, in practice, used in preparing rainfall estimates by this method.

Remote measurement of the after-effects of rainfall is based on detecting the changes in reflectance that occur as the soil gains water and, to a lesser extent, the changes in reflectance from vegetation. This is most successful in arid countries, where the bare ground is exposed, but at best it is qualitative, giving little more information other than that it has rained. The method cannot be placed in the same class as those using algorithms to quantify the amount of precipitation.

At the end of Chapter 8, the FRONTIERS project was introduced (Forecasting Rain Optimised using New Techniques of Interactively Enhanced Radar and Satellite data). This involves combining ground-based radar measurements of rainfall with estimates of the probability of rainfall based on satellite images. In the UK, where it was developed, FRONTIERS uses Meteosat data, comparing measured radar precipitation intensities with the reflectance and radiance values of the clouds deduced from visible and infrared data. Not only does it allow the radar measurements to have checks made on them (along with the additional input from ground-truth telemetering rain-gauges) but it allows the rainfall estimates to be extrapolated somewhat beyond the immediate radar coverage. As the name indicates, it is an interactive method and needs to involve an experienced forecaster who exercises judgement on how to modify the radar data, based on the satellite images, and produce an estimate of present rainfall along with a forecast of likely rainfall amounts over the next few hours. This combination of radar, telemetering ground-based raingauges and satellite data, as well as the input from the forecaster, produces an end product that is more than the sum of the parts. Nevertheless, while the satellite measurements do add useful extra information and will generally result in a better forecast, the satellite images could mislead and could have the opposite effect if wrongly interpreted.

Evaporation   Even with the best of ground-based instruments, it is difficult to measure evaporation, the eddy correlation method having an accuracy of only 15% to 25% (for hourly values). Most ground-based methods, such as that developed by Penman, estimate evaporation indirectly by measuring net radiation, temperature, humidity and wind speed. However, a simplified form of the Penman method (de Bruin 1987, Makkink 1957) uses measurements of incoming solar radiation (which can be measured by RS), combined with the saturation vapour pressure versus temperature curve (Chapter 4), using average monthly air temperatures for the area concerned. By using this abbreviated method, it is possible to estimate potential evaporation remotely from satellite data (Stewart *et al.* 1998b). It is, however, difficult to say just how accurate these estimates are.

Profilers

A new type of instrument, *a profiler*, has been developed recently which measures profiles of temperature, and wind speed and direction up to many kilometres altitude through the atmosphere by Doppler radar and Sodar (acoustic Doppler), which measures the shift in frequency of both radio and sound waves caused by wind, thermal structures and turbulence. Profilers do a similar job to radiosondes but can take continuous measurements, rather than just the one-shot of the sonde. However, they are expensive and are best suited to specialised research such as planetary boundary layer studies or for use at airports to detect wind shear and other dangers to aircraft such as downbursts.

References

Barrett, E. C. & Curtis, L. F. (1982) *Introduction to Environmental Remote Sensing,* second edition. Chapman & Hall, London, 352p.

Barrett, E. C. & Herschy, R. W. (1986) A European perspective on satellite remote sensing for hydrology and water management. Hydrologic Applications of Space Technology, IAHS Publ. No. 160, pp. 3–12.

Blyth, K. (1981) Remote sensing in hydrology. International Hydrological Report No. 74.

de Bruin, H. A. R. (1987) From Penman to Makkink. Evaporation and weather. In Volume 39, *Proc. and Information, TNO Committee on Hydrological Res., Technical Meeting 44.*

Carter, D. J. T. (1999) Variability and trends in the wave climate of the North Atlantic. *Proc. 9th Intl. Offshore and Polar Eng. Conf., Brest, France,* Volume III, pp. 12–18.

Dozier, J. (1987) Remote sensing of snow characteristics in southern Sierra Nevada. In *Proc. Vancouver Symp. on Large Scale Effects of Seasonal Snow Cover,* pp. 305–14. IAHS Publ. No. 166.

Dugdale, G. & Milford, J. R. (1986) Rainfall estimation over the Sahel using Meteosat thermal infrared data. In *Proc. Conf. ISLSCP,* ESA SP-248, pp. 315–19. European Space Agency, Paris.

Ebert, E. (1987) A pattern recognition technique for distinguishing surface and cloud types in polar regions. *J. Climate and Applied Meteorology,* **26,** pp. 1412–27.

Farnsworth, R. K., Barrett, E. C. & Dhanju, M. S. (1984) *Applications of remote sensing to hydrology including ground water.* International Hydrological Programme, UNESCO, 122p.

Harris, A. R. *et al.* (1998) Satellite retrievals of sea surface temperatures for long-term climate monitoring. In *Proc. Am. Met. Soc., UNESCO 25–29 May 1998,* Volume 1, pp. 162–5.

Harris, L. (1996) *Satellite Projects Handbook.* Newnes (Reed Elsevier), ISBN 0 7506 2406 X.

Jones, P. D. & Wigley, T. M. L. (1990) Satellite data under scrutiny. *Nature,* **344,** 711.

Makkink, G. F. (1957) Testing the Penman formula by means of lysimeters. *J. Inst. Water Eng.,* **11,** 277–88.

Möser, W. & Raschke, E. (1983) Mapping of solar radiation and of cloudiness from meteosat image data. *Meteorologische Rundschau,* **36,** 33–41.

Rango, A., Martinec, J., Foster, J. & Marks, D. (1983) Resolution on operational remote sensing of snow cover. Proc. Hydrological Applications of Remote Sensing and Remote Data Transmission, IAHS Publ. No. 145.

Schultz, G. A. & Barrett, E. C. (1989) Advances in remote sensing for hydrology and water resources management. UNESCO publ. IHP-III Project 5.1.

Spencer, R. W. & Christy, J. R. (1990) Precise monitoring of global temperature trends from satellites. Science, **247,** 1558–62.

Stewart, J. B, Engman, E. T., Feddes, R. A. & Kerr, Y. H. (1998) Scaling up in hydrology using remote sensing. *Int. J. Remote Sensing,* in press.

Stewart, J. B., Watts, C. J., Rodriguez, J. C., de Bruin, H. A. R., van den Berg, A. R. & Garatuza-payán (1999) Use of satellite data to estimate rainfall, radiation and evaporation for Northwest Mexico. Agricultural Water Management, **38,** 181–93.

Wong, K. W. (1975) Geometric and cartographic accuracy of ERST-1 imagery. *Photogrammetric Eng. and Remote Sensing,* **XLI,** No. 5.

15

Forward look

It has been shown throughout this book that most measurements of the natural environment are still being made using manual and mechanical instruments developed a century or more ago, albeit refined, but nevertheless limited. In consequence, we are less well informed about the environment than we would like to think, with data of uncertain accuracy and limited geographical coverage.

But with the developments of the past four decades, culminating in intelligent data loggers that record measurements from precise electronic sensors and are able, through satellite telemetry, to transmit their data from anywhere on earth, the environment can now be measured to much higher accuracy, with complete geographical coverage, in near-real time. This potential is only fully achieved, however, if the instruments are of good quality, correctly sited and well maintained. If they are not all these things, the data will be no better than those from the old instruments, possibly worse and certainly unreliable.

As in the past, so it will be in the future that the majority of measurements of the natural environment will continue to be made by individual organisations, small and large, commercial and governmental – research institutes, national weather services, water resources agencies and a host of others. These organisations will continue to buy, develop, operate and maintain equipment of their own choice, to their own standards and to suit their own budgets and purposes, just as they have in the past. But over the last decade, climate change has become a major scientific and political concern, and it is worth concluding this book with a brief look at how and why the climate should be better measured than it is being, or than it has been.

How well can we predict the future climate?

The Intergovernmental Panel on Climate Change (IPCC) was set up jointly by the WMO and the United Nations Environment Programme (UNEP) in 1988, its Working Group 1 being set the task of assessing available information on the science of climate change, in particular that arising from human activities. The basic physics of the 'greenhouse effect' is well understood and it is relatively easy to model, in isolation, the average global surface air temperature that would result from a given change in the level of atmospheric greenhouse gases (GHGs), such as carbon dioxide – provided that nothing else changed apart from the atmospheric temperature (Houghton 1997). Indeed, this exactly constitutes the IPCC 'projections' of future global warming and, for this reason, the IPCC does not call them climate 'predictions' because they are only theoretical 'explorations' of the possible effects of changing just the level of GHGs (Kattenberg *et al.* 1995).

Even the most sophisticated models are not yet able to predict what the real-world climate will do, for natural influences cannot yet be modelled with any certainty. The global circulation models (GCMs), as used in the IPCC assessments (Gates *et al.* 1995), although complex and state-of-the-art, requiring massive computing power, currently fall very far short in their ability to deal with the myriad of natural climatic processes. In particular, clouds and their radiative properties, the coupling between atmosphere and ocean, the hydrological balance, detailed processes at the land surface, ice and snow, biological processes in the sea, the land biosphere and the sources and sinks of greenhouse gases all operate at different time scales, magnitudes and locations, in ways yet barely understood. We are probably also still completely ignorant of many climate processes, as yet unforeseen. In addition, there are influences such as volcanic eruptions that are external to the climate system but which nevertheless affect it, and these cannot be modelled. The causes of the Little Ice Age, which lasted from about 1400 to 1850 AD, the Medieval Warm Period, at its peak between 1200 and 1300 AD, and of the Holocene maximum, which lasted from about 4000 to 3000 BC, are all quite unknown (Houghton 1997). Given such incomplete knowledge, we are not in any position to predict future climates and this makes it paramount that we measure what the climate actually does in reality. This in turn will also help to improve the models, and indeed the IPCC reports repeatedly stress the need for better data (Gates *et al.* 1995).

How well have we measured the recent past climate?

The main concern of the IPCC is that of 'global warming'. Assuming there is warming, and whatever its cause (entirely natural or in part anthropogenic – as yet we do not know which, or in what proportion), other changes, such as in precipitation, cloudiness, evaporation, humidity, soil moisture, river flow, wind, ice and snow cover, sea level, and numerous other factors, may well also occur as knock-on effects of any warming, although they would of course in turn probably then affect the warming that changed them, in a complex series of feedbacks. Because of the narrow terms of reference of the IPCC's Working Group 1 (the search for human influence on climate, principally through the burning of fossil fuels), the measurement of temperature is uppermost in their reports, with graphs showing anomalies of global annual mean air and sea-surface temperatures over the past 150 or so years (relative to a mean, either 1951–80 or 1961–90) (Nicholls 1995). However, from the preceding chapters it will be apparent that all measurements from the past, including temperature, as well as most measurements being made today, originate from simple, mechanical, manual instruments, with their manifest shortcomings (see Chapter 3, the paragraph headed mean temperatures and the section on exposure of thermometers). It comes as no surprise, therefore, that the data set the IPCC has been obliged to work with, in an attempt to establish how temperatures may have changed over the last century, is very far from perfect.

Much work has been required to 'homogenise' these data, largely air temperature measurements made by NWSs for weather forecasting (Jones *et al.* 1985, Parker 1994) and sea-surface temperatures from ships (Folland & Parker 1995, Parker *et al.* 1995), in attempts to correct their numerous errors. In the process, many records have had to be rejected as uncorrectable, uncheckable, or too short in duration, leaving 1584 out of 2666 land-surface air temperature stations in the northern hemisphere, mostly clumped together in the USA, Europe and the former USSR, and only 293 out of 610 in the southern hemisphere, concentrated in the southern countries of South America and in South Africa, Australia and New Zealand. Additionally, with 81% of the southern hemisphere being ocean, overall data coverage south of the equator is extremely sparse. Even in the northern oceans there are many data gaps, the oceans here accounting for 61% of the globe's surface. These measurements are the best we have, however, and the situation is not improving, with surface instrument networks being reduced in the poorer countries of Africa and Latin America.

The same situation applies to all the other climate variables, perhaps with even greater force, and indeed over the oceans we have virtually no measurements whatsoever of precipitation, evaporation, radiation or wind. So we are little better able to say with any certainty, from our instrument measurements, what the climate did in the past than we are able to predict, with models, what it might do in the future. However, as this book has demonstrated, we are most certainly in a position to measure, with much improved accuracy and geographical coverage, what the climate actually does in the future, should we so choose to make the investment needed.

Improving climate measurements for the future

In an attempt to bring some order to the collecting of hydrological, meteorological and oceanographic data, the WMO recently introduced its Global Climate Observing System (GCOS) (Kibby 1996) to co-ordinate data from the existing hotch potch of national and international instrument networks, one example of the latter being the World Hydrological Cycle Observing System (WHYCOS) (Rodda *et al.* 1993). However, the GCOS does not introduce any new instruments itself whatsoever, relying entirely on existing networks, over which it has no control; its success thus rests completely in the hands of the individual network operators, whose skill, commitment and permanence vary greatly from country to country and over time.

If we wish to obtain a reliable data set of climate variables in the future, it will be necessary to instal a new, worldwide, network of precision automatic instruments, dedicated to the task of climate monitoring, to a standard design throughout the world, independent of national networks, managed and inspected by an international team of experts; reliable measurements will then be obtained from selected variables at key sites, chosen for their scientific and geographical suitability rather than by historic chance and the dictates of shipping routes, as with the existing network. Remote sensing measures some variables better than *in situ* instruments ever can, but for others it is less well suited. The two paths will develop jointly, one assisting the other.

References

Folland, C. K. & Parker, D. E. (1995) Correction of instrumental biases in historical sea surface temperature data. *Quart. J. Roy. Met. Soc.*, **121**, 319–67.

Gates, W., Boer, G., Henderson-Sellers, A., Folland, C., Kitoh, A., McAvaney, B., Semazzi, F., Smith, N., Weaver, A. & Zeng, Q. (1995) Climate models – Evaluation. In *Climate Change 1995. The Science of Climate Change, Contribution of Working Group 1 to the Second Assessment Report of the IPCC*, pp. 229–84. Cambridge University Press, ISBN 0 521 56436 0.

Houghton, J. T. (1997) *Global Warming: The Complete Briefing*, second edition. Cambridge University Press. ISBN 0 521 62932 2.

Jones, P. D., Raper, S. C. B., Santer, B., Cherry, B. S. G., Goodes, C., Kelly, P. M., Wigley, T. M. L., Bradley, R. S. & Diaz, H. F. (1985) The grid point surface air temperature data set for the Northern Hemisphere. US Department of Energy, Report TRO22.

Kattenberg, A., Giorgi, F., Grassl, H., Meehl, G., Mitchell, J., Stoufer, R., Tokioka, T., Weaver, A. & Wigley, T. (1995) Climate models – Projections of future climate. In *Climate Change 1995. The Science of Climate Change, Contribution of Working Group 1 to the Second Assessment Report of the IPCC*, pp. 285–357. Cambridge University Press, ISBN 0 521 56436 0.

Kibby, H. (1996) The global climate observing system. *WMO Bull.*, **45**, 140–7.

Nicholls, N., Gruza, G. V., Jouzel, J., Karl, T. R., Ogallo, L. A. & Parker, D. E. (1995) Observed climate variability and change. In *Climate Change 1995. The Science of Climate Change, Contribution of Working Group 1 to the Second Assessment Report of the IPCC*, pp. 133–92. Cambridge University Press, ISBN 0 521 56436 0.

Parker, D. E. (1994) Effects of changing exposure of thermometers at land stations. *Int. J. Climatology*, **14**, 1–31.

Parker, D. E., Folland, C. K. & Jackson, M. (1995) Marine surface temperature: observed variations and data requirements. *Climatic Change*, **31**, 559–600.

Rodda, J. C., Pieyns, S. A. & Matthews, G. (1993) Towards a world hydrological cycle observing system. *J. Hydrol. Sci.*, **38**, 373–8.

Appendix: Acronyms

ACCLAIM	Antarctic Circumpolar Current Level from Altimetry and Island Measurements
ADC	analogue-to-digital conversion
APT	automatic picture transmission
AWS	automatic weather station
BAS	British Antarctic Survey
BCD	binary coded decimal
BOD	biochemical oxygen demand
BSI	British Standards Institute
CLS	Collecte Localisation Satellite
CMOS	complementary metal-oxide semiconductor
CNES	Centre National d'Études Spatiales (French Space Agency)
CRAWS	cold regions automatic weather station
DCP	data collection platform
DFIR	double-fence international reference (snow fence)
DO	dissolved oxygen
EEPROM	electrically erasable and programmable read only memory
EGOS	European Group on Ocean Stations
EPROM	electrically programmable read only memory
ESA	European Space Agency
GCM	Global circulation model
GCOS	Global Climate Observing System
GMS	Geostationary Meteorological Satellite
GOES	Geostationary Operational Environmental Satellite
GOMS	Geostationary Operational Meteorological Satellite
GPS	global positioning system
GTS	global telecommunication system
HF	high frequency

IC	integrated circuit
IH	Institute of Hydrology
IOC	Intergovernmental Oceanographic Commission
IPCC	Intergovernmental Panel on Climate Change
IPTS	international practical temperature scale
IR	infrared
JMA	Japan Meteorological Agency
LED	light emitting diode
LES	land earth station
LORAN	long-range navigation
LRIT	low-rate information transmission
LRPT	low-rate picture transmission
LSB	least significant bit
MDWF	moisture dry weight fraction
MOS	metal-oxide semiconductor
MSB	most significant bit
MSS	multispectral scanner
MVF	moisture volume fraction
MVP	moisture volume percentage
NASA	National Aeronautical and Space Administration
NEDIS	National Environmental Satellite, Data and Information Service
NOAA	National Oceanographic and Atmospheric Administration
NPL	National Physical Laboratory
NWS	national weather service
ORP	oxidation–reduction potential
OWS	ocean weather ships
PAR	photosynthetically active radiation
PC	personal computer
PDUS	primary data user system
PRT	platinum resistance thermometer
PSTN	public switched telephone network
PTFE	polytetrafluoroethylene
RAM	random access memory
RH	relative humidity
RMS	Royal Meteorological Society
RS	remote sensing
SAR	synthetic aperture radar
SCADA	system control and data acquisition
SDUS	secondary data user system

SST	sea-surface temperature
SVP	saturation vapour pressure
TDR	time domain reflectrometry
TM	thematic mapper
TOGA	Tropical Ocean Global Atmosphere
UHF	ultra-high frequency
UV	ultraviolet
VHF	very high frequency
VMO	Voluntary Marine Observer
VOF	Voluntary Observing Fleet
VP	vapour pressure
VWC	volumetric water content
WHYCOS	world hydrological cycle observing system
WMO	World Meteorological Organisation
WWW	World Weather Watch
ZFP	zero-flux plane

There are some groups of letters that look like acronyms (e.g. Meteosat, Inmarsat, Wefax, Landsat) but they are in fact abbreviations, although often wrongly printed in upper-case. Their meaning can usually be inferred from the context in which they are quoted and so they are not included above. However, a useful list of related acronyms and similar abbreviations has been compiled by Padgham (1992), amounting to around 3000 items.

Reference

Padgham, R. C. (1992) A directory of acronyms, abbreviations and initialisms. Natural Environment Research Council, Polaris House, Swindon, UK.

Index

DATE DUE

DEMCO 13829810